ChemActivities

and LabActivities

for General Education Chemistry

Christopher Nichols
California State University, Chico

Kendall Hunt
publishing company

Kendall Hunt
publishing company

www.kendallhunt.com
Send all inquiries to:
4050 Westmark Drive
Dubuque, IA 52004-1840

Printed in the United States of America
10 9 8 7 6 5 4 3 2

ISBN 978-1-5249-2684-7

Contents

Section C: LabActivities

Lecture 1:
CHEM 100
Syllabus Highlights
and Introduction

Christopher J. Nichols

What is chemistry?

- The study of matter, and its composition, properties, and changes

化学

change - science

- The study of the elements
- An element is a fundamental substance which cannot be broken down into simpler substances
- Modern theory describes each element as having a unique number of protons in its atomic nucleus. The concept of an element (ancient) predates the atomic theory (1808)

Why is chemistry so hard?

- Atoms are SMALL: you can't see them:
- http://www.youtube.com/watch?v=x6AJWMPNSr0
- (jump to 6:25)
- How small?
- http://www.youtube.com/watch?v=9JJHC5y_RKE
- (jump to 2:45)

But we *can* actually 'see' them

With atomic force microscopy...

Lecture 2

Elements, Compounds, and Mixtures

Pure substances: Elements

- Copper
  - $Cu_{(s)}$
- Oxygen
  - $O_{2\,(g)}$

Pure substances: Compounds

- Water
  - $H_2O_{(l)}$
- Salt
  - $NaCl_{(s)}$

Mixtures: Homogeneous

- Seawater
  - H_2O, NaCl, other ions, microorganisms, etc.
- Hard candy
  - $C_{12}H_{22}O_{11}$, H_2O, $C_{18}H_{14}N_2Na_2O_8S_2$
- Brass
  - Cu, Zn
- 14K gold
  - Au, Ag
- Air
  - N_2, O_2, Ar, CO_2, H_2O, other gases

Mixtures: Heterogeneous

- Mud
  - Water, sand
- Vinaigrette
  - Oil, vinegar
- Me
  - All sorts of things; mostly H_2O

What about…

- Jell-O?
- Yellow-cake Uranium?
- Uranium (IV) oxide?
- Milk?
- Mercury?
- Tomato juice?
- Cling wrap?

Lecture 3:
Nuclear Power and Nuclear Weapons

CHEM 100

Chemical Reactions vs. Nuclear Reactions

In a CHEMICAL reaction, atoms in compounds or molecules can be rearranged to form other molecules, but the atoms themselves remain unchanged:

e.g. The combustion of octane (in gasoline):

$2\ C_8H_{18} + 25\ O_2 \longrightarrow 16\ CO_2 + 18\ H_2O$

Chemical Reactions vs. Nuclear Reactions

In a NUCLEAR reaction, the nature of the atoms themselves can change: actually altering the numbers of protons and neutrons in each individual atom's nucleus:

e.g. The fission of Uranium-235

$^{235}U + {}^{1}n \longrightarrow {}^{142}Ba + {}^{91}Kr + 3\ {}^{1}n$

Energy Differences: Chemical vs. Nuclear

Reaction	Type	Heat released
Detonation of TNT: $2\ C_7H_5N_3O_6 \rightarrow 3\ N_2 + 5\ H_2O + 7\ CO + 7\ C$	Chemical	5 kJ per g
Combustion of octane: $2\ C_8H_{18} + 25\ O_2 \rightarrow 16\ CO_2 + 18\ H_2O$	Chemical	48 kJ per g
Fission of uranium $^{235}U + {}^{1}n \rightarrow {}^{142}Ba + {}^{91}Kr + 3\ {}^{1}n$	Nuclear	71,000,000 kJ per g
Fusion of hydrogen (in the sun) $4\ {}^{1}H \rightarrow {}^{4}He + 2\ e^{+}$	Nuclear	600,000,000 kJ per g

Getting to Grandma's House:

If I drive to Grandma's house (62 miles away) in my wife's Toyota, I use about 3.1 gallons of gas (around 8.2 kg):

If I could design an engine which would use the other reactions as efficiently I could get there with:

- 78 kg of TNT ... but this might post a problem
- ... or 5.5 mg of uranium for fission
- ... or 0.7 mg of hydrogen for fusion

Nuclear Power for transportation

- There are vehicles which use uranium for propulsion:

- There aren't any vehicles that use hydrogen fusion for propulsion: why not? ...
- You need a temperature of 40,000,000 ˚C for the reaction to occur.

Controlled Nuclear Reactions: Nuclear Power Plants

- Uranium is found in nature as two major isotopes:
  - U-238 (99.3% but not readily fissionable)
  - U-235 (0.7% and undergoes fission, but has low abundance)
- To make the fission reaction more efficient, the fuel pellets must be ENRICHED to 3-5% U-235.

The Manufacture and use of Enriched Uranium

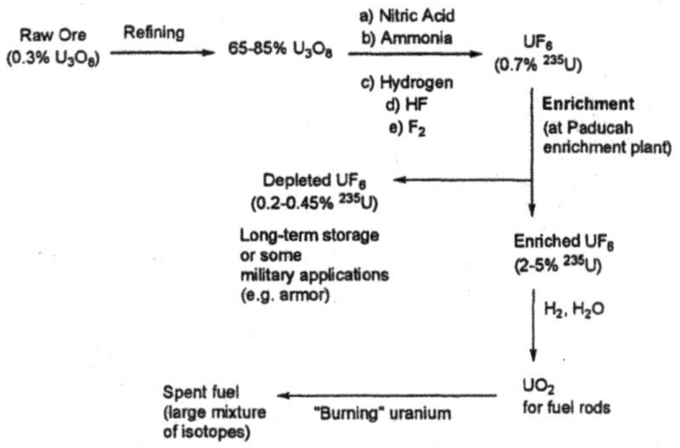

Uranium Enrichment:

Natural Uranium		Low-Enriched Uranium		Highly Enriched Uranium
0.7% ^{235}U	→	2-5% ^{235}U (for power plants) 12-19% ^{235}U (for research reactors)	→	85% (weapons-grade)

Uranium Power Plants

Pros and Cons of Nuclear Energy

- Large amount of power produced for small amount of fuel
- No greenhouse gas emissions
- Coal mining is dangerous

- Spent nuclear fuel
- Depleted UF_6
- Accidents (Chernobyl)

Nuclear Weapons: Chain Reaction

- Since fission can be started by a neutron, and since more neutrons are produced in each fission event, it is possible for a **chain reaction** to take place
- If you have a **critical mass** of ^{235}U (33 lbs) you have enough fuel to sustain a chain reaction, and if the mass can be held together long enough, it can explode

"Little Boy" (dropped on Hiroshima)

Explosive power = 13-16 kilotons of TNT (ca. 30 million lbs) from approx. 140 lbs of uranium.

Thermonuclear Weapons

- A ^{235}U or ^{239}Pu fission explosion is initiated:
- This triggers a 2nd-stage fusion reaction with much more power:
- $^{6}Li + {}^{1}n \rightarrow {}^{3}H + {}^{4}He$
- **$^{3}H + {}^{2}H \rightarrow {}^{4}He + {}^{1}n$**

Castle Bravo explosion: March 1954
(largest test detonation by the USA):
Yield = 15 megatons

Tsar Bomba nuclear test: October 1961
(largest test detonation by the USSR):
Yield = 50-57 megatons

Lecture 4: Big and Small

Numbers we're used to:

88 miles to Sacramento airport
157 lbs on my bathroom scale
29' 2½" is the long jump world record
$5.79 for a box of Cheerios

Numbers we're not so good at

BIG NUMBERS
5,878,625,373,196 miles (1 light-year)
602,214,000,000,000,000,000,000 (Avogadro)

SMALL NUMBERS
0.000000000000000000000000000000910939 kg
(mass of an electron)
0.00000000007414 meters (H-H bond length)

How to get a grip on these?

Scientific Notation

$10^1 = 10$

$10^2 = 10 \times 10 = 100$

$10^3 = 10 \times 10 \times 10 = 1000$

$10^{12} = 10 \times 10 \ldots = 1{,}000{,}000{,}000{,}000$ (12 zeros)

$5.88 \times 10^{12} = 5{,}880{,}000{,}000{,}000$

So 1 light year = 5.88×10^{12} miles

Your turn...

- 602,214,000,000,000,000,000,000 (Avogadro's number)
- $26,060,000,000

Getting a grip part 2

Scientific Notation for small numbers

$10^{-1} = 1 \div 10 = 0.1$

$10^{-2} = 1 \div (10 \times 10) = 0.01$

$10^{-3} = 1 \div (10 \times 10 \times 10) = 0.001$

$10^{-11} = 1 \div (10 \times 10 \ldots) = 0.00000000001$ (10 zeros after the decimal)

$7.414 \times 10^{-11} = 0.00000000007414$ meters

So H-H bond is 7.414×10^{-11} meters long (short)

Your turn…

- 0.000000000000000000000000000000910939 kg
- 0.0000000000088 mL

A big problem with chemistry:

- Some things are too SMALL
  - Atoms
  - Electrons
  - Molecules
- Other things are too BIG
  - Numbers of atoms or molecules in visible amounts of material

Dealing with big and small

- Different Units for Different Sizes:
  - The Metric System

Length: meter   Mass: gram   Volume: liter

Making the units bigger and smaller with prefixes

- kilo (k) = 1000x (10^3)
- mega (M) = 10^6 x
- giga (G) = 10^9 x
- centi (c) = 0.01x (10^{-2})
- milli (m) = 0.001x (10^{-3})
- micro (μ) = 10^{-6} x
- nano (n) = 10^{-9} x

There are other prefixes (from 10^{24} to 10^{-24}) but aren't used as often as those above.

(see the on-line textbook, Chapter 1, section 1.4)

Dealing with big and small

A new unit called the MOLE:

1. A pigmented spot, mark, or small permanent protuberance on the human body
2. Any of numerous burrowing insectivores
3. A spy (as a double agent) who establishes a cover long before beginning espionage
4. An abnormal mass in the uterus (med.)
5. A massive work of masonry and large stones or earth laid in the sea as a pier or breakwater
6. A spicy sauce made with chilies and usually chocolate and served with meat

Dealing with big and small

- A new unit called the MOLE
  - ~~Official definition: a mole is the number of items in exactly 12 grams of ^{12}C atoms.~~
  - Practical definition: a mole of anything is 6.02 x 10^{23} items of that thing.
  - One mole of atoms = 6.02 x 10^{23} atoms.

"Avogadro's Number"

Counting atoms (1)

If atoms weren't so small we could count them one by one:

...but they are, and we can't.

Counting atoms (2)

- One way to measure the number of atoms we have in a sample of an element is to weigh it on a balance:

2 gold coins = 62.2 grams of pure gold.

Counting atoms (3)

"62.2 grams" of gold doesn't tell us how many gold atoms there are

unless...

we have a conversion factor that will let us relate the number of atoms to their mass...

WE DO!

MOLAR MASS

Look at the periodic table:

79	← 79 protons per atom
Au	← Tells you it's gold
196.9665	← **Atomic mass in grams per mole**

In other words: one mole of gold atoms has a mass of 196.9655 grams

Or you can say that:
Gold has a **molar mass** of

$$\frac{196.9655 \text{ grams}}{1 \text{ mole of gold atoms}} \quad \text{or: } 196.9655\ \frac{g}{mol}$$

So how many gold atoms were there in those two gold coins? Well - let's figure out how many moles first...

Your turn

Who is using a pencil to write things down?

For each letter you write, you are using about 7 μg of graphite (carbon) from your pencil. How many carbon atoms is that?

What's in the Water?

CHEM 100
Week 5

Would you drink this water…

If I told you it contained:

- Arsenic?
- Radioactive alpha-emitters?
- Vanadium?
- Tetrachloroethylene?
- A herbicide?

All of these have been detected in Chico's water supply.

You should ask me:

How much was there? And how much is bad for me?

No water is 100.0000% pure…

<u>Contaminants (regulated by EPA)</u>

- Microorganisms
- Disinfectants (Cl_2 and by-products)
- Inorganic chemicals
- Organic molecules
- Radioactive elements

What's in... Seawater?

(total 3.5% salinity)

Ion	Concentration (ppm)	Concentration (mol/L)
Cl^-	19,345	0.546
Na^+	10,752	0.468
SO_4^-	2,701	0.028
Mg^{2+}	1,295	0.053
Ca^{2+}	416	0.010
K^+	390	0.010
HCO_3^-	145	0.002

What is "ppm"?

Normally reported by mass:
1 ppm means that there is 1 unit of mass of solute for every 1,000,000 units of mass of solution

Example: 1 ppm = 1 mg per kg (1 mg per liter of water)
Or: 1 ppm = 1 lb in 1,000,000 lbs (500 tons)
What does 1 ppm look like?

1 **ppb** means that there is 1 unit of mass of solute for every 1,000,000,000 units of mass of solution

Example: 1 ppb = 1 mg per 1,000 kg (1 mg per 1 m^3 water)

Mineral Content: Sea, tap, bottled

Source	Ca^{2+} (ppm)	Mg^{2+} (ppm)	Na^+ (ppm)
Ocean	416	1295	10,752
Tap (SJ)	56	14	57
Tap (Phx)	51	29	169
Tap (avg)GW	34	10	35
Evian	78	24	5
Perrier	147	3	9
Arrowhead	20	5	3
MRI (daily)	1000 mg	300-400 mg	<3000 mg

GW = groundwater   1 ppm = 1 mg/L

Maximum Concentration Levels (MCL) - set by the EPA

These levels are determined by:

- National Drinking Water Advisory Council (NDWAC)
- Representatives from water utilities
- Environmental groups
- Public interest groups
- States, tribes and the general public
- Minority and low-income communities
- Representatives of small businesses

Maximum Concentration Levels (MCL) - set by the EPA

Contaminant	Max level (ppm)	Max level (ppb)
Arsenic (As)	0.010	10
Cadmium (Cd)	0.005	5
Chromium (Cr)	0.1	100
Cyanide (CN^-)	0.2	200
Fluoride (F^-)	4.0	4000
Lead (Pb)	0.015	15
Mercury (Hg)	0.002	2

Contaminant content

Source	As (ppb)	Pb (ppb)	Hg (ppb)
MCL	**10 ppb**	**15 ppb**	**2 ppb**
Ocean	2.6 ppb	0.03 ppb	0.15 ppb
Chico Tap	0-2 ppb	0	0
Laytonville	50 ppb	0	0
Stockton	4.5 ppb	0-13 ppb*	0-2 ppb*
Arrowhead	0-3.2 ppb	-	-
Crystal Geyser	0-4.8 ppb	-	-
Volvic	12-14 ppb	-	-

Human Toxicity

- Arsenic
  - Skin lesions, neuropathy, anemia, lung caner, skin cancer
- Lead (chronic exposure)
  - Children show weight loss, weakness, anemia
- Mercury
  - Inflammation of mouth and gums, loose teeth, kidney damage, tremors, nervousness

Organic Contaminants

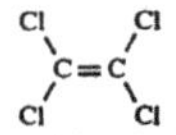

Tetrachloroethylene (Perchloroethylene, or PERC) MCL = 5 ppb

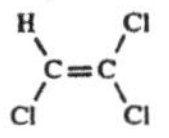

Trichloroethylene (TCE) MCL = 5 ppb

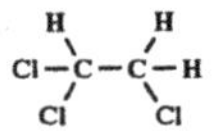

1,1,2-Trichloroethane MCL = 5 ppb

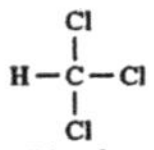

Chloroform

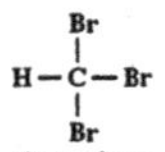

Bromoform

MCL = 100 ppb (combined)

Chico has had problems with this:

- Link to DTSC report for Hegen Lane
- Link to DTSC report for Central plume
- Chico tap water: of 208 tests from 1998-2003, 10 came back positive for PERC (max level 9.5 ppb)

Is bottled water any better?

- Source: NRDC Tests
- Most bottled water samples had no organic contamination

So why do we drink so much bottled water, then?

Lecture 6: Water Purification

CHEM 100

Some water purification methods

- Distillation
- Flocculation/Sedimentation
- Ion Exchange
- Filtration
- Reverse Osmosis

Which is a microscopic view of water boiling?

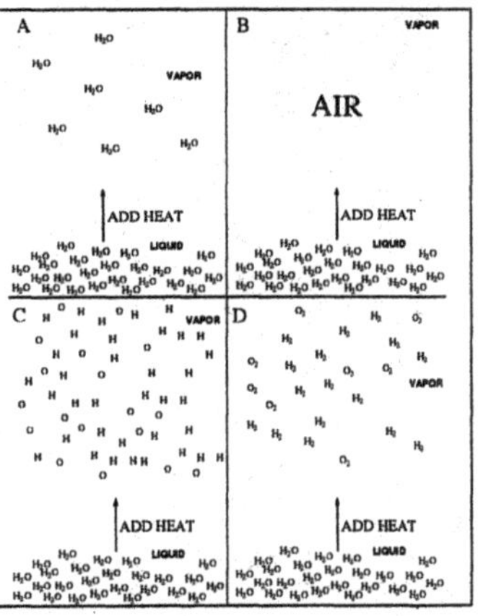

One method of water purification: DISTILLATION

$$H_2O_{(l)} \xrightarrow[\text{add heat}]{} H_2O_{(g)} \xrightarrow[\text{remove heat}]{} H_2O_{(l)}$$

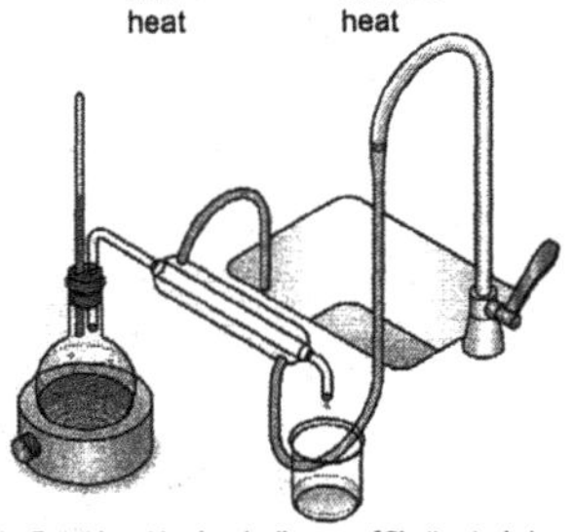

Issues with Distillation

- Water has a high boiling point (100 °C) - takes a lot of heat energy to take water at 'room temperature' (25 °C) up to 100 °C.
- Water also has a high heat of vaporization - takes a lot of energy to get liquid water at 100 °C to turn into water vapor (at 100 °C)
- Where does the energy come from? Probably burning fuel...

$$H_2O_{(l)} \xrightarrow[\text{add heat}]{} H_2O_{(g)} \xrightarrow[\text{remove heat}]{} H_2O_{(l)}$$

Distillation on large scale (MSF)

In MSF, you don't need near as much heat: the heat 'lost' as the brine condenses is transferred to incoming seawater to start warming it up.

Produces 85% of the world's desalinated water in places like...

Saudi Arabia United Arab Emirates

Australia (Perth)

Flocculation/Sedimentation

According to the IUPAC definition, flocculation is "a process of contact and adhesion whereby the particles of a dispersion form larger-size clusters." (thanks Wikipedia)

Small particles form a slurry

When they aggregate into larger ones, they sink.

The industrial water can then be re-used...

Flocculants are anionic polymers

Very large molecules made of repeating units of a monomer

Polyacrylamide

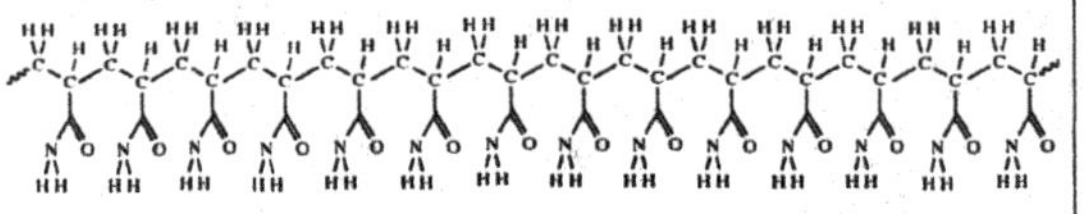

Polyacrylamide co-polymerized with acrylic acid (anionic)

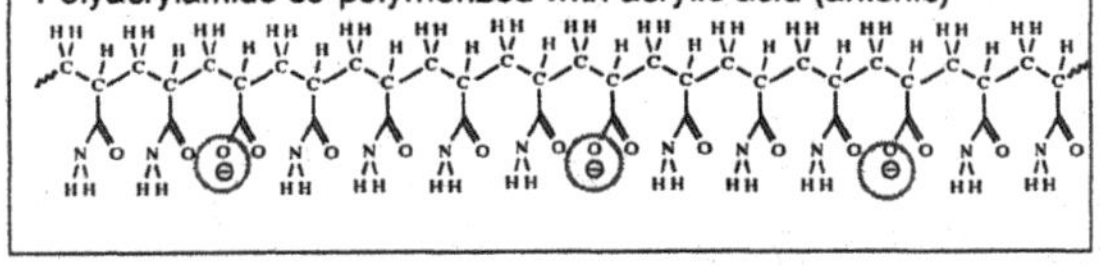

SNF Inc. - polymer manufacturer

http://www.snfinc.com/canadaoilsands.html

Ion exchange: to reduce water hardness

http://en.wikipedia.org/wiki/Water_softening#Effects_of_sodium

Filtration

Many refrigerator filters have three systems:

Charcoal (catches organic impurities)

Fibers (catch large particles and microorganisms)

Ion-exchange (lowers water hardness, catches heavy metals)

Osmosis

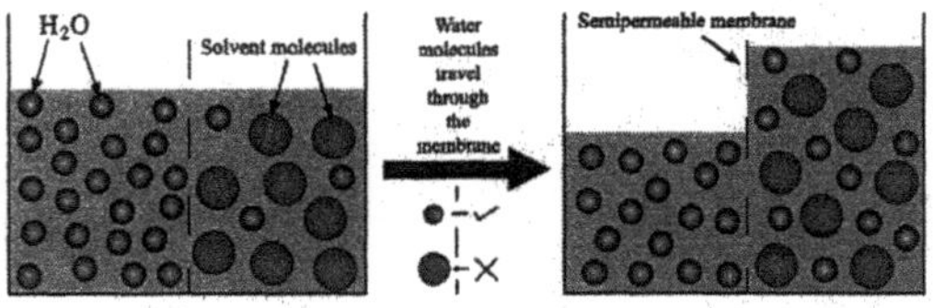

© 2013 by Dreamy Girl. Used under license of Shutterstock, Inc.

Reverse Osmosis: large scale

http://www.scientificamerican.com/article.cfm?id=california-desalination-reverse-osmosis

Sand City, CA	250,000 gallons per day
Hadera seawater reverse osmosis plant (Israel)	120 million gallons fresh water per day (max) 50,000+ membrane units

Issues with Reverse Osmosis

- Energy needed to maintain water pressure (pumps)
- Membranes don't last forever ($$)
- Pre-treatment needed to extend membrane lifetime

What is Science?

... and what isn't science?
Chem 100 - Lecture 7

What Science Is

- Science is an intellectual activity carried out by humans that is designed to discover information about the natural world in which humans live and to discover the ways in which this information can be organized into meaningful patterns.
- A primary aim of science is to collect facts (data).
- An ultimate purpose of science is to discern the order that exists between and amongst the various facts.

From Dr. Sheldon Gottlieb in a lecture series
at the University of South Alabama: April 3, 1997

The Scientific Method (simplified)

1. Gather Data
2. Form Hypothesis
3. Test Hypothesis
4. Results of tests confirm, refute, or modify hypotheses, which lead to gathering of further data and more testing

Example: Rutherford's Gold Foil Experiment

Some opinions about Science

- I think that we shall have to get accustomed to the idea that we must not look upon science as a "body of knowledge", but rather as a system of hypotheses, or as a system of guesses or anticipations that in principle cannot be justified, but with which we work *as long as they stand up to tests*, and of which we are never justified in saying that we know they are "true" . . . (Karl R. Popper (1902-1994))
- We [scientists] wouldn't know truth if it jumped up and bit us in the ass. We're probably fairly good at recognizing what's false, and that's what science does on a day-to-day basis, but we can't claim to identify truth. (Dr. Steven M. Holland, University of Georgia Geology Professor)

What is a **THEORY**?

- In normal conversational use: **theory** means *a guess* or *a hunch*

When scientists say "theory"

- A well-substantiated explanation of some aspect of the natural world; an organized system of accepted knowledge that applies in a variety of circumstances to explain a specific set of phenomena
- A scientifically accepted general principle or body of principles offered to explain natural phenomena that is consistent with evidence, data, and experimental results. Theories can be disproved, but it is scientifically impossible to prove a theory correct.
- An accepted collection of ideas, explanations, and testable hypotheses describing a broad, important set of phenomena. The acceptance arises from repeated successful testing of a theory's many hypotheses.

Examples of major scientific theories

- Einstein's theory of relativity
- Quantum mechanical theory
- Atomic theory
- Theory of Natural Selection

Characteristics of a scientific theory

1. A simple unifying idea that postulates nothing unnecessary ("Occam's Razor")
2. Logically consistent
3. Logically and empirically falsifiable (i.e., cases must exist in which the theory can be imagined to be invalid)
4. Clearly limited by explicit boundary conditions so that it is clear whether or not particular data are or are not relevant to verification or falsification
5. Be empirically testable or lead to predictions or retrodictions that are testable
6. Actually make verified predictions and/or retrodictions
7. Involve reproducible results
8. Provide criteria for the interpretation of data as factual, artifactual, anomalous or irrelevant

(From "Lance F.", from atheism.about.com)

Evolution: Theory or Fact?

- Debate over definitions:
- In science "fact" can only mean "confirmed to such a degree that it would be perverse to withhold provisional consent." I suppose that apples might start to rise tomorrow, but the possibility does not merit equal time in physics classrooms.

Stephen J. Gould, "Evolution as Fact and Theory"; Discover, May 1981

How to tell if a claim is scientific

- Reproducibility
- Ability to make predictions (or retrodictions)
  - Gallium & Germanium
  - Eclipses that happened in the past
- Testability and Falsifiability
- Adequate supporting evidence?
  - Anecdotal evidence insufficient
  - Personal experience insufficient
- Includes or suggests a mechanism
  - e.g. Remote Viewing

What science cannot do

- Make moral judgments (i.e. HESC research)
- Make aesthetic judgments (Beethoven v. Mozart)
- Answer questions about the 'supernatural'
- Prove a negative (logically impossible)
- Find "the truth"
- Deal with issues of human consciousness and feelings… *for now!*

Some Scientific Revolutions (in chronological order)

- Shape of the Earth
- Relative Position of the Earth
- Age of the Earth
- Origin of humans and other species
- Change in configuration of continents
- Origin of the Universe

Lecture 8: Scientific Revolutions
CHEM 100

For each of the 6 ideas we'll look at:

1. Pre-scientific view:
2. Scientific view:
3. Evidence:
4. First proponents:
5. Implication for humans:
6. 'Traditionalist' response:
7. View of people today:

Shape of the Earth

1. Pre-scientific view: Flat
2. Scientific view: Almost spherical
3. Evidence: Receding objects descending into horizon, Magellan, Photos from space
4. First proponents: Ancient Greeks
5. Implication for humans: Not much
6. 'Traditionalist' response: Very few still object
7. View of people today: Accept scientific view

Relative Position of Earth

1. Pre-scientific view: Center of the Universe (Sun orbits Earth)
2. Scientific view: Earth orbits sun (and sun isn't special)
3. Evidence: Astronomical observations with telescopes
4. First proponents: Copernicus (1540), Galileo (1630)
5. Implication for humans: People are not at the 'center of the universe' either
6. 'Traditionalist' response: Copernicus's book called 'hypothetical' in preface; Galileo imprisoned and forced to recant
7. View of people today: Earth orbits sun

Age of Earth

1. Pre-scientific view: Young (ca. 6000 years)
2. Scientific view: 4.6 billion years
3. Evidence: Rates of geologic formation; radiometric dating of rocks
4. First proponents: Hutton (1790s), Lyell (1840-1870)
5. Implication for humans: Recent inhabitants of very old planet: earth was without humans for much of its history.
6. 'Traditionalist' response: Young Earth Creationism
7. View of people today: Some accept scientific view, others do not.

Origins of humans (and other species)

1. Pre-scientific view: Divine creation
2. Scientific view: Evolution from earlier life forms
3. Evidence: Fossils, structural similarities between humans and primates, DNA evidence
4. First proponents: Darwin, Wallace, Huxley (1800s)
5. Implication for humans: Descended from 'lesser' non-human species, therefore not 'special'
6. 'Traditionalist' response: Heated debate in 1800s; focus of 20th and 21st century creationism
7. View of people today: Split between scientific and non-scientific: heated debate continues including in legal system.

The Evolution Controversy

Let me try to make crystal clear what is established beyond reasonable doubt, and what needs further study, about evolution. **Evolution as a process that has always gone on in the history of the earth can be doubted only by those who are ignorant of the evidence or are resistant to evidence, owing to emotional blocks or to plain bigotry.** By contrast, the mechanisms that bring evolution about certainly need study and clarification. There are no alternatives to evolution as history that can withstand critical examination. Yet we are constantly learning new and important facts about evolutionary mechanisms.

Theodosius Dobzhansky (1973)

The Evolution Controversy

R. C. Lewontin "Evolution/Creation Debate: A Time for Truth" Bioscience 31, 559 (1981)

Movement of Continents

1. Pre-scientific view: Static Geography
2. Scientific view: Movement via plate tectonics
3. Evidence: Geology of continents, age of sea floor, earthquakes, volcanoes, mountains
4. First proponents: Wegner, Wilson, Hess (early 1900s)
5. Implication for humans: Not much
6. 'Traditionalist' response: virtually none
7. View of people today: Don't care

Origin of the Universe

1. Pre-scientific view: Divine creation
2. Scientific view: "Big Bang"
3. Evidence: Red shift of light from distant galaxies, cosmic microwave background radiation
4. First proponents: Hubble (early-mid 1900s)
5. Implication for humans: Earth is young relative to the universe; universe existed without Earth or humans for billions of years
6. 'Traditionalist' response: Creationism
7. View of people today: Some accept the scientific view, some reject it.

Conclusion

- Science hasn't been kind to humanity's notion of itself: but then it's not supposed to be. That can lead to reactionary attitudes…
  "…. Man was not an accident and reflects an image and likeness unique in the created order. Those aspects of evolutionary theory compatible with this truth are a welcome addition to human knowledge...." (Sam Brownback, 5/31/07)
- Scientific theories are well-substantiated explanations of facts from observations of nature.
- Science starts with data, and develops theories based on the data: not the other way around.
- Science can't answer all questions, but then it doesn't pretend to.

CHEM 100: Lecture 10
Radiation and Global Warming

Questions you should be able to answer:

- 1. What is the source of the vast majority of the earth's energy?
- 2. What types of electromagnetic radiation are there?
- 3. What types of radiation does the Sun send our way? How does it get here?
- 4. Do all of the types of solar radiation make it to the surface of the Earth? For those that don't, why don't they?
- 5. How do molecules absorb different types of electromagnetic radiation?
  - a. ultraviolet
  - b. visible (scattering)
  - c. infrared
  - d. microwave
- 6. What is heat?
- 7. How does the greenhouse effect work?

Earth's energy comes from…

… the Sun

The sun's radiation is primarily:

- Ultraviolet light
  - 250-400 nm wavelength; high-energy
- Visible light
  - 400-700 nm wavelength
- Infrared light
  - 700-2500 nm wavelength; low-energy

Some solar radiation gets blocked by the atmosphere

UV-A:

- 320-400 nm wavelength
- Lower energy
- Atmosphere lets UV-A through
- Responsible for tanning
- Increased risk for skin cancer

UV-B:

- 290-320 nm wavelength
- Higher energy
- 99% of it blocked by ozone layer
- Responsible for sunburn
- Increased risk for carcinoma

Visible light:

- 80% of it reaches the surface
- Scattered by gas molecules in air (that's why the sky is blue)

Why is some IR radiation absorbed by the gases in the air?

- Lewis structures make it look like the molecules are "balls and sticks"

H-O-H O=C=O

- In fact the molecules are vibrating

IR spectrum of $CO_{2\ (g)}$

not absorbed

strongly absorbed

3333 nm (3.3 µm) ← wavelength → 20,000 nm (20 µm)

IR spectrum of $H_2O_{\ (g)}$

not absorbed

strongly absorbed

2500 nm (2.5 µm) ← wavelength → 20,000 nm (20 µm)

Where does 'heat' come into this?

- Thermal energy is kinetic energy of random molecular motion, and is measured by its temperature:
  - Atoms/molecules of high-temperature substances are moving more rapidly
  - Atoms/molecules of low-temperature substances are moving more slowly
- Heat is the amount of thermal energy transferred when a cold object contacts a hot object.
- Heat can also be transferred by radiation (without coming into thermal contact)
- As heat is added to a system, its molecular motion increases; that can cause a rise in temperature or a phase change.

What makes H_2O and CO_2 'greenhouse gases'?

- Light energy (UV, Vis, IR) from the sun enters earth's atmosphere. Some is absorbed, but most of it gets to the surface, which absorbs it.
- The surface re-emits the radiation, primarily as IR radiation. Some of that is radiated back out to space, but most of it is absorbed by the atmosphere.
- The more IR radiation absorbed by the atmosphere, the more energy it contains, so the higher its temperature will be.
- H_2O and CO_2 are effective at absorbing this IR radiation: the more of it is present in the atmosphere, the more IR radiation it absorbs, and the higher its temperature gets.

Warming of the atmosphere by greenhouse gases

- More CO_2 in the air leads to increased energy absorption and a higher average temperature
- CO_2 in air: 280 ppm (~1800); 385 ppm (2008)

What if there were no greenhouse gases?

- Earth's actual average surface temperature: 14 °C
- Earth's estimated average surface temperature without a greenhouse effect: –18 °C
- Venus's actual average surface temperature: 462 °C
- Venus's estimated average surface temperature without a greenhouse effect: –34 °C (colder than Earth, despite being closer to the sun, since its surface is more reflective)

Lecture 11: Reactions and Energy

CHEM 100

Energy can be used to:

- Warm things up
- Do "work"
- Cause phase changes

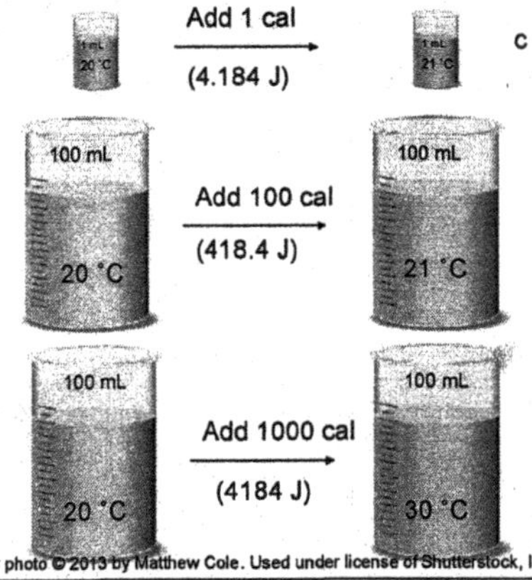

$$c = \frac{1\ \text{cal}}{\text{g}\cdot{}^{\circ}\text{C}}$$

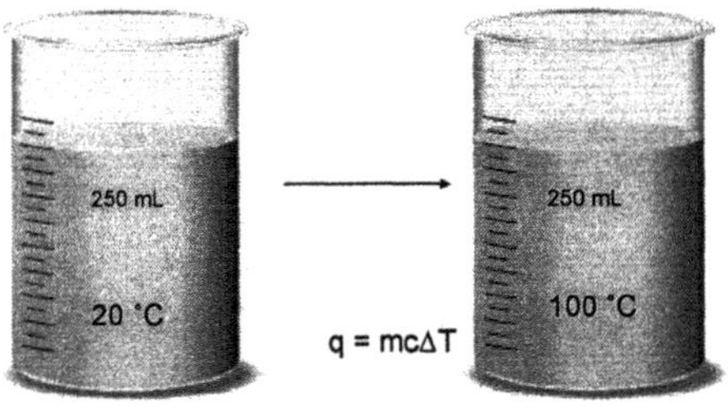

q = mass X temp.change X 'heat capacity'

$$250\text{ g} \times 80\text{ °C} \times \frac{1\text{ cal}}{\text{g}\cdot\text{°C}} = 20{,}000\text{ cal} = 20\text{ Cal (food)}$$

$$20{,}000\text{ cal} \times \frac{4.184\text{ J}}{1\text{ cal}} = 83{,}680\text{ J}$$

Which of the following is true?

1. Breaking a covalent bond always requires energy
2. Breaking a covalent bond sometimes requires energy and sometimes releases energy
3. Breaking a covalent bond always releases energy

$$\text{H—H} \longrightarrow \text{H}^{\bullet} + {}^{\bullet}\text{H}$$

$$\text{H-H} + 436\text{ kJ/mol} \longrightarrow \text{H} + \text{H}$$

$$\text{O=O} + 498\text{ kJ/mol} \longrightarrow \text{O} + \text{O}$$

$$\text{O} + \text{H} \longrightarrow \text{O-H} + 467\text{ kJ/mol}$$

$$2\ H_2 + O_2 \longrightarrow 2\ H_2O$$

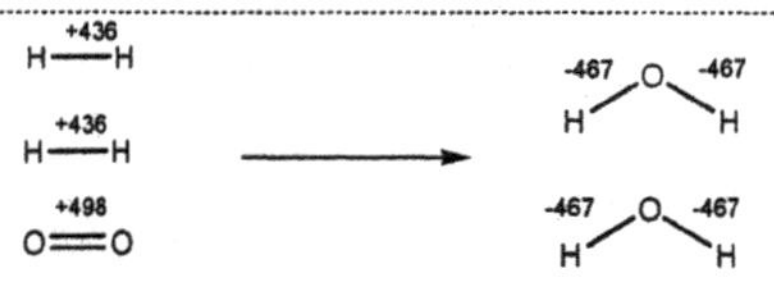

ENERGY IN: 1370 kJ   ENERGY OUT: 1868 kJ

NET RESULT: 498 kJ released.

Different fuels release different amounts of energy (per unit mass)

Fuel	Energy content
Wood	14 kJ per gram
Ethanol (C_2H_6O)	30 kJ per gram
Coal (C)	31 kJ per gram
Gasoline (mostly C_8H_{18})	48 kJ per gram
Propane (C_3H_8)	51 kJ per gram
Methane (CH_4)	56 kJ per gram
Hydrogen (H_2)	140 kJ per gram

Joules are cheap...

How many Joules of energy can I get from $1 of gas?

$$\$1 \times \frac{1 \text{ gall}}{\$3.12} \times \frac{3785 \text{ mL}}{1 \text{ gall}} \times \frac{0.74 \text{ g}}{1 \text{ mL}} \times \frac{48{,}000 \text{ J}}{1 \text{ g}}$$

= 43,000,000 J

Lecture 13: Fuels and Alternative Fuels

CHEM 100

Topics for today

- Generation of Electricity
- Primary sources of energy: fossil fuels
  - Coal
  - Oil
  - Gas
- Carbon Cycle
- Renewable sources of energy:
  - Conventional (hydro, nuclear)
  - Alternative (biomass, wind, geothermal, solar

The basics of electricity generation

- A fuel is burned to boil water, converting it to high-pressure steam
- As the steam expands and cools, it pushes the fans of a turbine, which makes it spin
- The turbine's shaft is connected to a coil of wire within a magnetic field (a 'dynamo'). This generates an alternating current (AC) of electricity
- The steam condenses back to water (its heat removed by a secondary coolant), and is pumped back into the boiler.

US Energy Use, 2009

	Quadrillion BTU	%
TOTAL	94.6	100.0%
Fossil Fuels	78.5	83.0%
Oil	35.4	37.4%
Gas	23.4	24.7%
Coal	19.7	20.8%
Nuclear	8.4	8.9%
Renewables	7.6	8.0%
Biomass	3.9	4.1%
Hydroelectric	2.7	2.9%
Wind	0.7	0.7%
Geothermal	0.2	0.2%
Solar	0.1	0.1%

Source: www.eia.gov

Oil Refining

- Start with 1 barrel (approx. 42 gallons) of crude oil and you get:
  - 19.6 gallons of gasoline
  - 10.0 gallons of diesel and heating oil
  - 4.0 gallons of jet fuel
  - 1.7 gallons of heavy fuel oil
  - 1.7 gallons of natural gas (LRG)
  - 7.6 gallons of 'other'

Hydraulic fracturing ('Fracking'):

- A well is dug deep into shale rock (horizontally in many cases)
- Fracturing fluid is pumped into the rock at high pressure to induce fractures into the rock
  - Fluid is 85% water, 13% sand, 1.3% HCl, and small amounts of other components
  - The fluid flows back and must be treated before disposal or re-use
- The fissures allow the natural gas to escape out the well at an increased rate, making it economically viable.

Carbon 'footprint' for electricity generation

Fuel	Lbs CO_2 released per kWh of electricity generated
Coal	2.02 - 2.12
Oil	1.59 - 1.70
Natural Gas	1.12
Hydroelectric	0
Nuclear	0
Wind or Solar	0
Biomass	Non-zero

Source: www.eia.gov

Biomass fuels (US consumption in 2009)

- Wood (1.9 QBTU)
  - $(C_6H_{10}O_5)_n + 6n\ O_2 \rightarrow 6n\ CO_2 + 5n\ H_2O$ + heat
- Ethanol (0.9 QBTU)
  - Produced by fermentation of sugar:
    - $C_6H_{12}O_6 \rightarrow 2\ CO_2 + 2\ C_2H_5OH$
  - Burned to produce energy:
    - $C_2H_5OH + 3\ O_2 \rightarrow 2\ CO_2 + 3\ H_2O$
  - An additional 0.6 QBTU of non-renewable energy is needed to produce this ethanol.**

Biodiesel (0.04 QBTU in 2009)

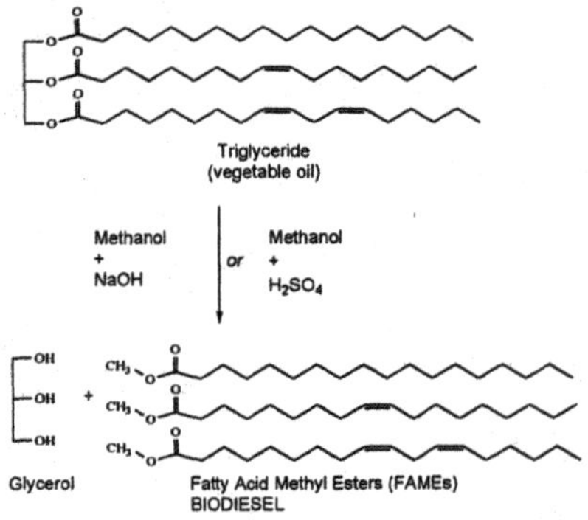

Why are biomass fuels classified differently from fossil fuels when they both produce CO_2 upon combustion?

Other alternative fuel sources:

- Wind
  - 0.2 QBTU in 2005; 0.7 QBTU in 2009
- Solar Thermal and Solar Photovoltaic (PV)
  - 0.1 QBTU in 2005 and 2009
- Currently: very minor contributors to US energy consumption (less than 1%)

Lecture 14:
Organic Chemistry

CHEM 100

Two definitions of "organic"

- Organic food: food produced without the use of synthetic pesticides or fertilizers. It also typically means the food was produced without antibiotics, additives, and there was no genetic modification.
- Organic chemistry: chemistry of compounds with carbon (C) as the central atom.

- Are Doritos "organic"?

Organic vs. Inorganic compounds

- Obviously compounds without C are inorganic:
  - NaCl
  - SiO_2
  - PCl_5
  - KOH
- Most compounds with C are organic:
  - CH_4
  - C_2H_5OH (ethanol)
  - $C_6H_{12}O_6$ (sugar)
  - $C_{17}H_{21}NO_4$ (cocaine)
- Some carbon-containing compounds are still inorganic:
  - CO_2
  - $CaCO_3$ (limestone)
  - $NaHCO_3$

CARBON MAKES 4 BONDS

Drawing organic molecules

- Expanded Form
- Condensed Form (seen mostly when the formula is typed in a line of text)
- Zig-zag shorthand
  - At the end of a line is a C (unless the line is pointing to another atom)
  - At each bend is a C
  - H's are not shown unless they're on an atom besides C; they're assumed to be there such that each C has 4 bonds.

Organic Compounds can be classified by their functional group

- Hydrocarbons (contain C and H only)
  - Alkanes (all C-C single bonds)
  - Alkenes (C=C double bonds)
  - Alkyne (C-C triple bonds)
  - Cycloalkanes (C's form a ring)
  - Aromatic compounds (include a benzene ring)

cyclohexane (C_6H_{12}) benzene (C_6H_6)

More functional groups

- Oxygen-containing functional groups
  - Alcohols (C-O-H)
  - Ethers (C-O-C)
  - Ketones and Aldehydes
  - Carboxylic acids (like acetic acid; vinegar)
  - Esters (like lipids)

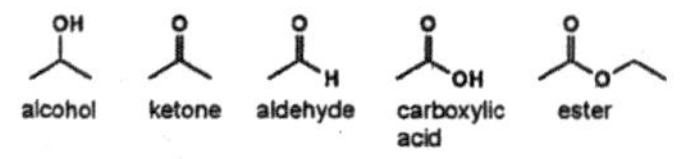

Yet more functional groups

- Nitrogen-containing groups
  - Amines and amides
- Multi-functional compounds like:
  - Sugars (alcohols and ethers)
  - Proteins (amides plus several others)
  - RNA (aromatics, alcohols, phosphates, etc)

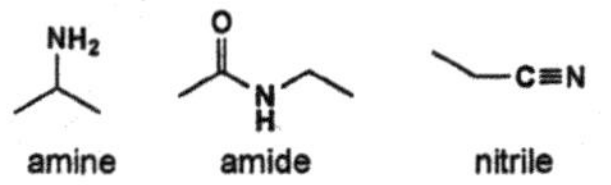

Many organic compounds contain functional groups with POLAR bonds

- Each atom has a specific 'pull' on electrons, called electronegativity:
- O = 3.5
- N = 3.0
- C = 2.5
- H = 2.1
- If a covalent bond is between two atoms with very different electronegativities, the electrons in the covalent bond are not equally shared.

Lecture 15:
Food and Drug Chemistry

Some Basic Foodstuffs

A triglyceride (fat)

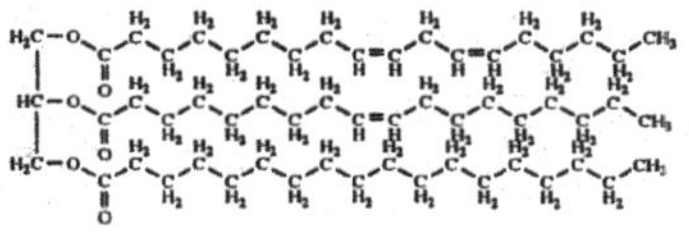

A triglyceride (fat) written in organic chemistry shorthand

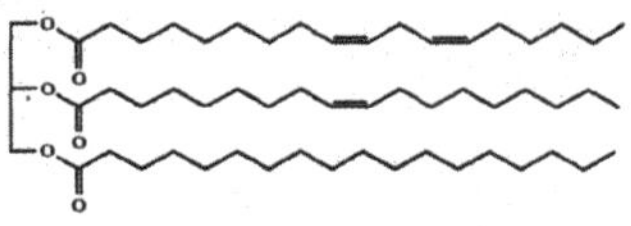

More Basic Foodstuffs

Amino Acid

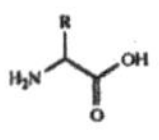

Protein

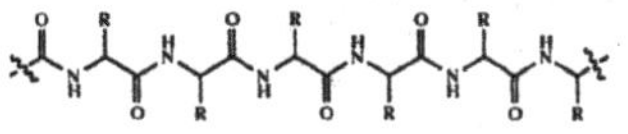

Glucose (a sugar)

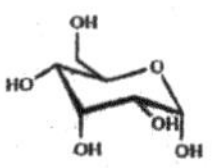

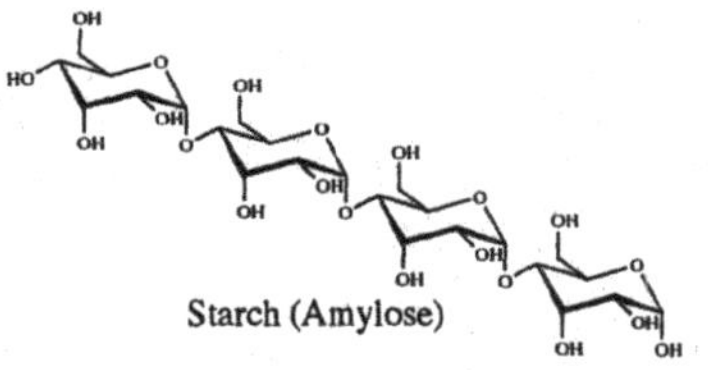

Starch (Amylose)

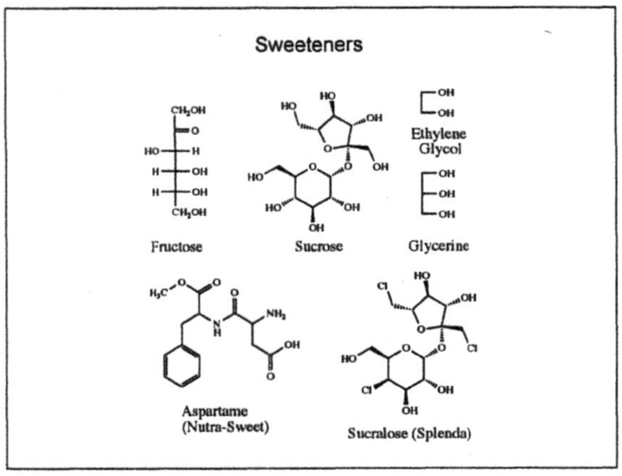
Sweeteners
Fructose
Sucrose
Ethylene Glycol
Glycerine
Aspartame (Nutra-Sweet)
Sucralose (Splenda)

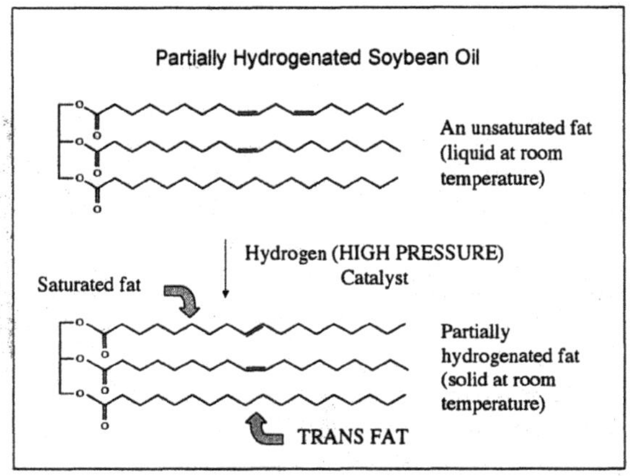
Partially Hydrogenated Soybean Oil
An unsaturated fat (liquid at room temperature)
Hydrogen (HIGH PRESSURE) Catalyst
Saturated fat
Partially hydrogenated fat (solid at room temperature)
TRANS FAT

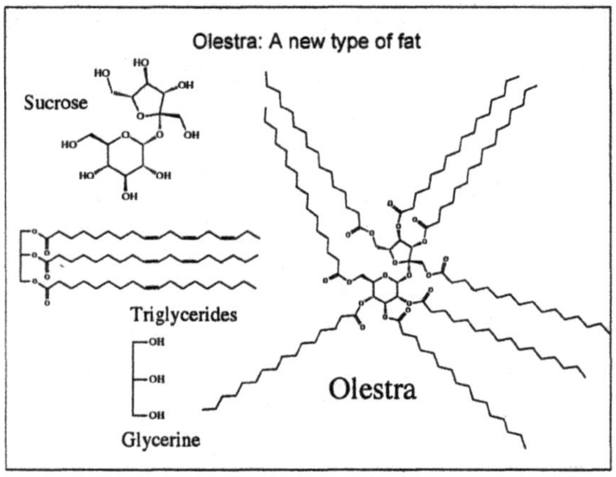
Olestra: A new type of fat
Sucrose
Triglycerides
Glycerine
Olestra

DRUGS

What is organic chemistry really all about?

What organic chemistry is all about

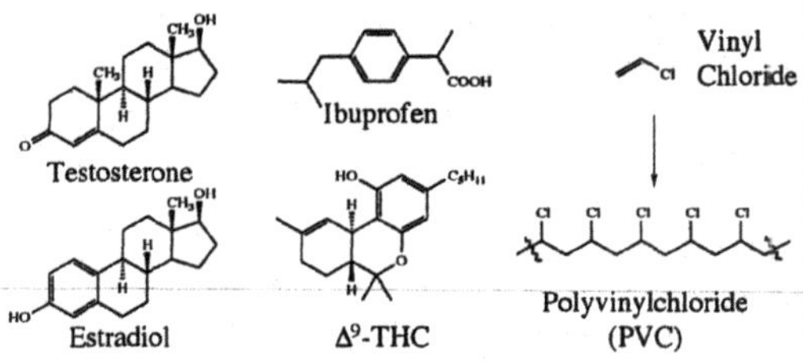

Top 10 Drugs in USA: 1983

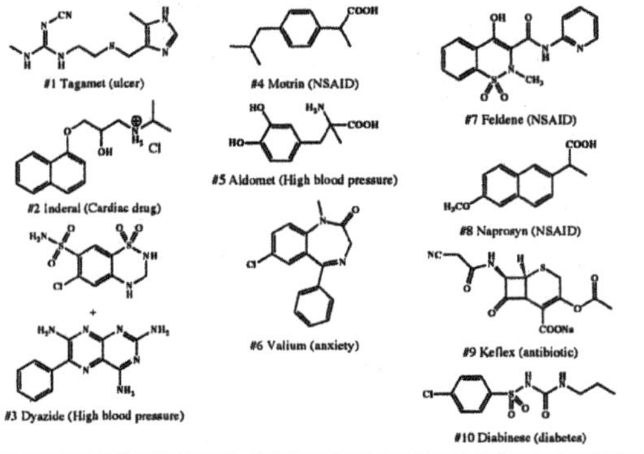

Top 10 Prescription Drugs (2006)

(by US # of prescriptions)

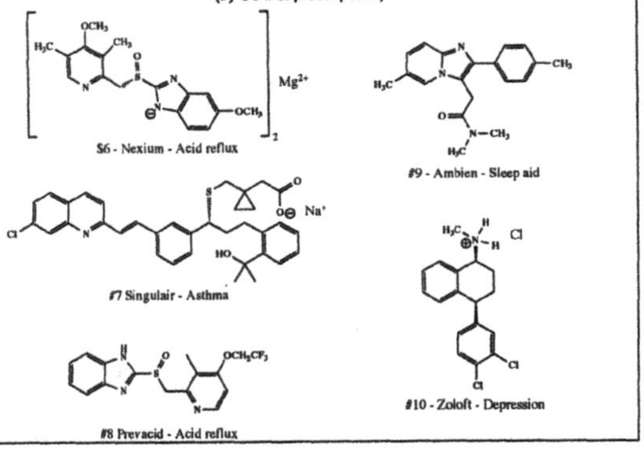

Top 10 Prescription Drugs (2006)

(by US # of prescriptions)

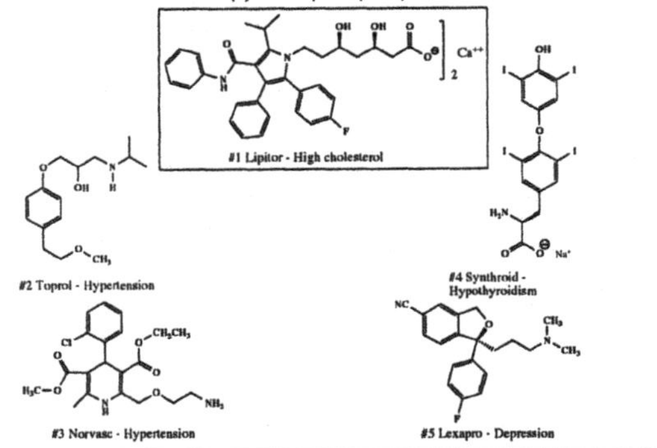

Birth Control Pills

Ortho-Novum 7-7-7

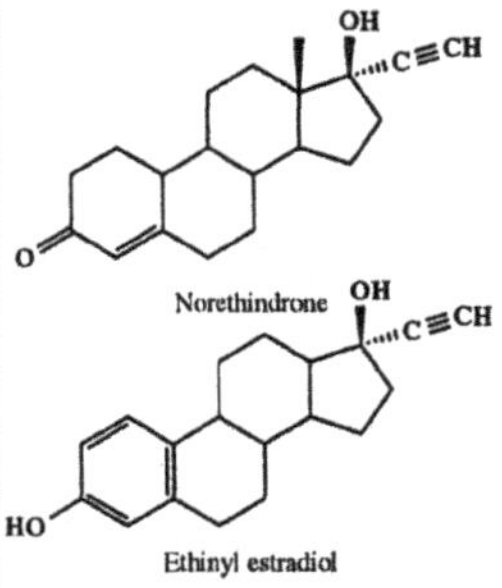

White	Light Peach	Peach
0.5 mg	0.75 mg	1.0 mg
0.035 mg	0.035 mg	0.035 mg

Total monthly supply of drugs: 16.5 mg (!)

Alkaloids

Naturally-occurring nitrogen-containing compounds found in plants

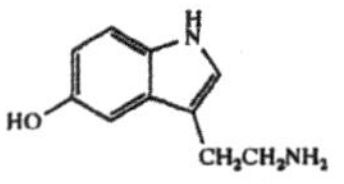

Serotonin

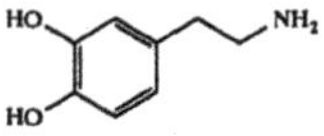

Dopamine

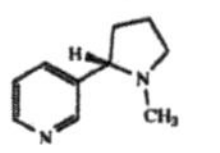

Nicotine

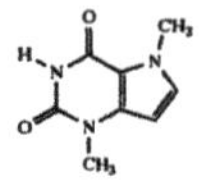

Theobromine (chocolate)

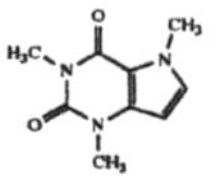

Caffeine

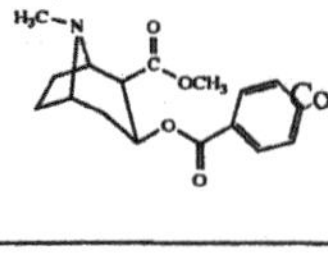

Cocaine

Illegal Drugs

Who uses them?

Survey (2004): "In the past month have you used any illicit drug?"

Any illicit drug includes marijuana/hashish, cocaine (including crack), heroin, hallucinogens (including LSD and PCP), inhalants, or any prescription-type psychotherapeutic drug used nonmedically.

Age	Use
12-13 years	3.8%
14-15 years	10.9%
16-17 years	17.3%
18-25 years	19.4%
26-34 years	11.1%
35 years +	4.2%

The most popular illegal drugs

Incidence of use by high school seniors, 2005

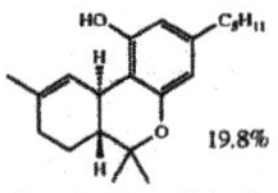

19.8%

Active ingredient in marijuana/hashish (Δ^9-THC)

2.0%

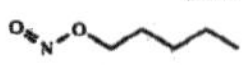

One example of an inhalant (amyl nitrite)

2.3%

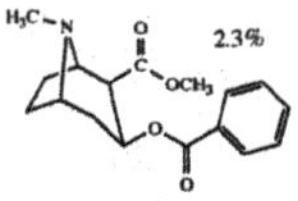

Cocaine

1.0%

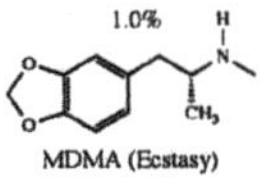

MDMA (Ecstasy)

Other illegal drugs

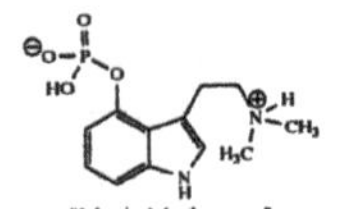

"Magic Mushrooms" contain psilocybin

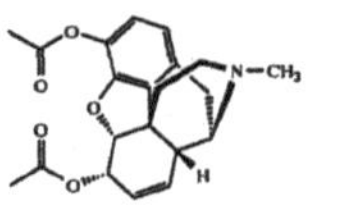

heroin

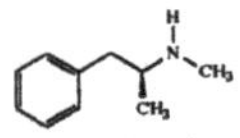

Methamphetamine

"Beavis & Butt-Head" synthesis of Methamphetamine

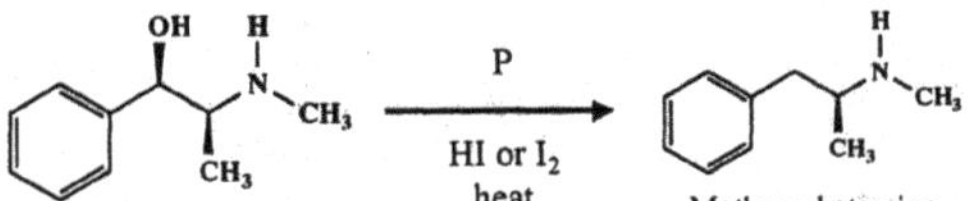

Sudafed → Methamphetamine

Sudafed: obtained from cold medicines

P: obtained from matchbook striking strips

I_2: not hard to get… mixing P and I_2 makes HI anyway

Food and Drug Chemistry

CHEMISTRY 100: ChemActivity 1**
Atomic Structure

Model 1: Schematic diagram for various atoms and ions

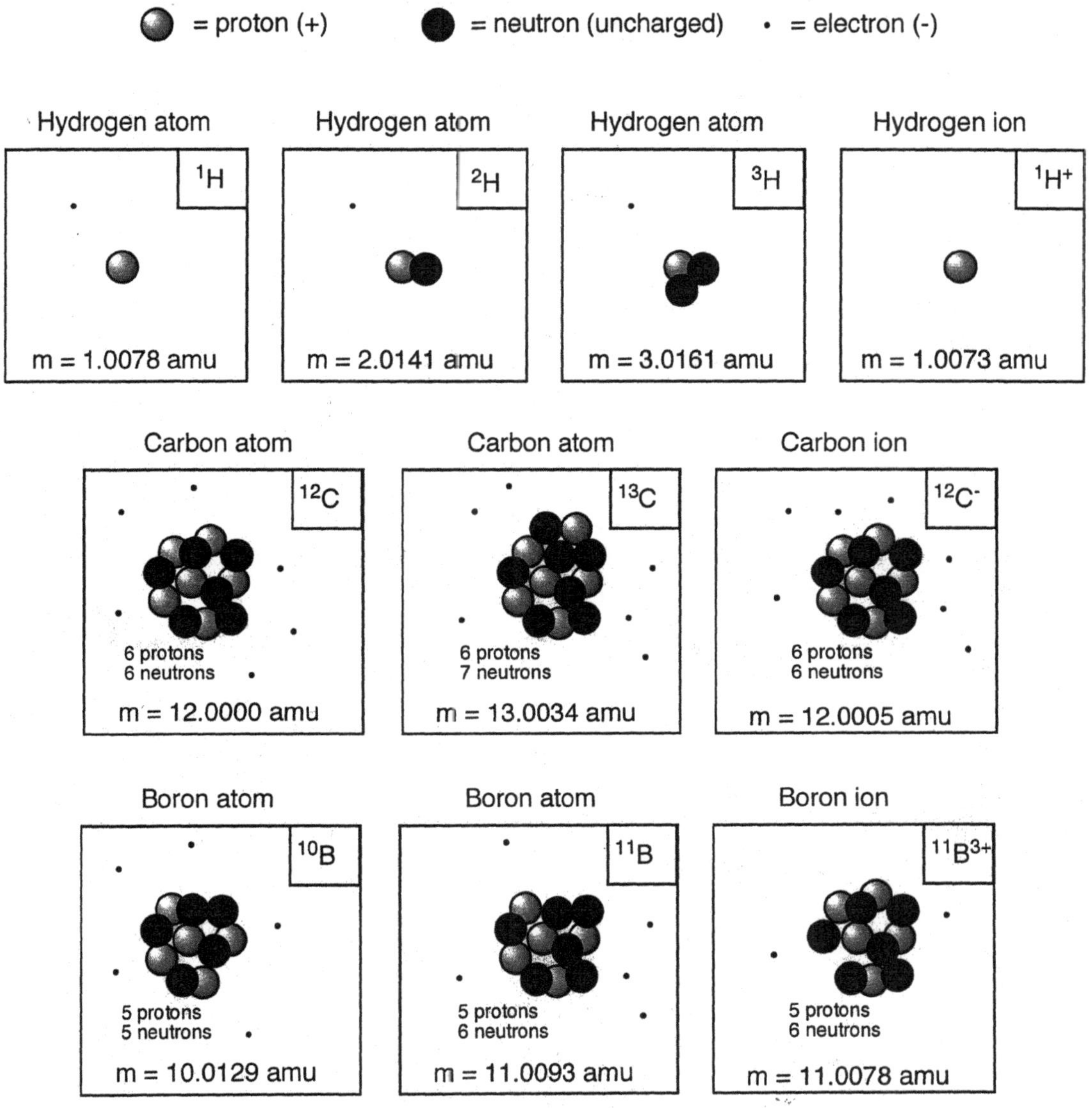

- ^{1}H, ^{2}H, and ^{3}H are **isotopes** of hydrogen.
- ^{12}C and ^{13}C are **isotopes** of carbon.
- ^{12}C is pronounced "Carbon-12"
- The protons and neutrons are found in the atomic **nucleus**.

** Adapted from ChemActivity 3, Garoutte, M. P. *General, Organic, and Biological Chemistry: A Guided Inquiry*, 2007.

Group Name ____________________ Recorder Name __________________

Critical Thinking Questions

1.a) How many protons are present in ^{12}C? _____ ^{13}C? _____ $^{12}C^{-}$? _____

b) How many neutrons are present in ^{12}C? _____ ^{13}C? _____ $^{12}C^{-}$? _____

c) How many electrons are present in ^{12}C? _____ ^{13}C? _____ $^{12}C^{-}$? _____

d) What do all carbon atoms and ions have in common?

2. What do all of the hydrogen atoms/ions have in common?

3. What do all the boron atoms/ions have in common?

4. Look at a periodic table (there is one in your folder). Above each element is its **atomic number**. Based on your answers to the previous questions, how does the atomic number of an element relate to its composition?

5. How many protons does a chlorine atom (Cl) have? ______ If you added one more proton to a chlorine atom, would it still be chlorine? Explain.

6. Below each schematic diagram is the mass of that atom/ion (given in amu, a very small unit of mass). Look at the diagrams for the hydrogen atoms and ions and fill in this table:

Atom / Ion	# of protons	# of neutrons	# of electrons	mass (rounded off to nearest whole number)
^{1}H				
^{2}H				
^{3}H				
$^{1}H^{+}$				

Using the table above, calculate the mass (to the nearest whole number of amu's) of:

A proton ______ A neutron ______ An electron ________

7. Does an electron have any mass at all? Explain.

8. The superscript next to the elemental symbol (like the "2" in ^{2}H) is called the **mass number**.

a) How does the mass number relate to the mass (in amu) of the atom/ion?

b) Determine a relationship between the mass number and the number of protons, neutrons, and/or electrons in an atom/ion.

c) Does your answer to Question 8b agree with your answers to Question 6? Explain how.

9. Some of the species are listed as ‘ions’ rather than ‘atoms’.

a) What differences do you notice in the elemental symbols of atoms and ions?

b) What differences do you notice in the number of protons, neutrons, and/or electrons between atoms and ions of hydrogen?

c) What differences do you notice in the number of protons, neutrons, and/or electrons between atoms and ions of the other elements?

d) Explain the significance of the superscripts + and - and ‘3+’ in the elemental symbol. How does that symbol relate to the numbers of protons, neutrons, and/or electrons in the ion?

Exercises:

Reading Assignment: please read Section 3.4 of the online textbook http://preparatorychemistry.com/Bishop_Atoms_First.htm for more information.

1. Draw a structural drawing (akin to those shown in Model 1) for the following species:

^{7}Li $^{19}F^{-}$ $^{24}Mg^{2+}$

2. True/False Statements: circle the correct answer

a)	Every atom of aluminum (Al) will have 13 protons	T	F
b)	Every atom of aluminum will have 14 neutrons	T	F
c)	Every atom of aluminum will have 13 electrons	T	F
d)	Every ion of aluminum will **not** have 13 electrons	T	F

3. Explain your answers to Exercises 2(c) and 2(d).

(2c)

(2d)

4. Fill in all the blanks in the table below:

Symbol	**Atomic Number**	**Mass Number**	**Number of Protons**	**Number of Neutrons**	**Number of Electrons**	**Atom or Ion?**
^{40}K						
			16	16	16	
$^{60}Ni^{2+}$						
	53	127			54	

CHEMISTRY 100: ChemActivity 2
Molecules and Chemical Reactions

Group Name ____________________ Recorder Name __________________

Chemistry wouldn't be much fun if you only dealt with static compounds and elements. The heart of chemistry is *transformation*: in fact the Chinese and Japanese representation of the word chemistry is 化学, where 化 means "change" and 学 means "science" – chemistry is indeed "the science of change". Today we will study the representation of these changes as balanced chemical equations.

Model 1: The formation of water, written two ways:

Using balls to represent atoms: Using chemical formulas:

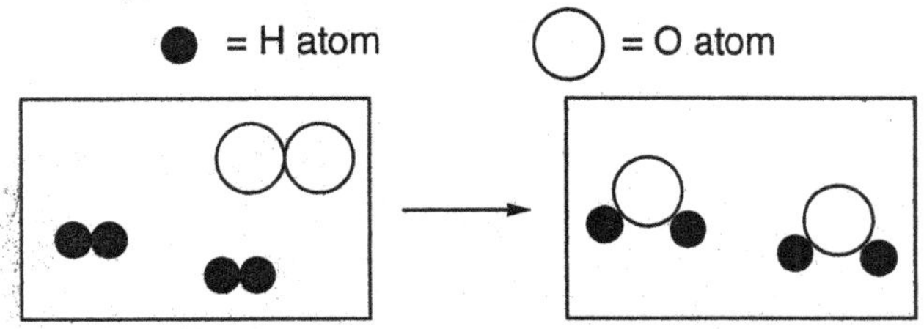

or $2\ H_2 + O_2 \rightarrow 2\ H_2O$

Critical Thinking Questions:

1 a) This diagram  represents a hydrogen **molecule**. How is its chemical formula written?

1 b) This diagram  represents an oxygen **molecule**. How is its chemical formula written?

1 c) This diagram  represents a water **molecule**. How is its chemical formula written?

2. Write one sentence which explains the relationship between <u>atoms</u> and <u>molecules</u>.

3. What do the subscript numbers (*i.e.* the "2" in H_2O) within each molecule's formula represent?

4. In the reaction, what do the large numbers (*i.e.* the first '2' in "**2** H_2O") in front of each compound's formula represent? (*Note: these numbers are called the* ***coefficients***)

5. What does it mean if there is no coefficient in front of a compound's formula in a reaction?

6. In either representation, what are the total number of H atoms:
   a) on the left side of the arrow? ______
   b) on the right side of the arrow? ______

7. What are the total number of O atoms:
   a) on the left side of the arrow? ______
   b) on the right side of the arrow? ______

8. What are the total number of molecules:
   a) on the left side of the arrow? ______
   b) on the right side of the arrow? ______

9. Based on your answers to questions 6, 7, and 8, what do you predict stays the same in the course of a chemical reaction and what can change?

Model 2: The fermentation of sucrose, represented two ways

This reaction is carried out by yeasts in the production of alcoholic beverages, the rising of bread dough, and in the production of ethanol from corn for use as fuel.

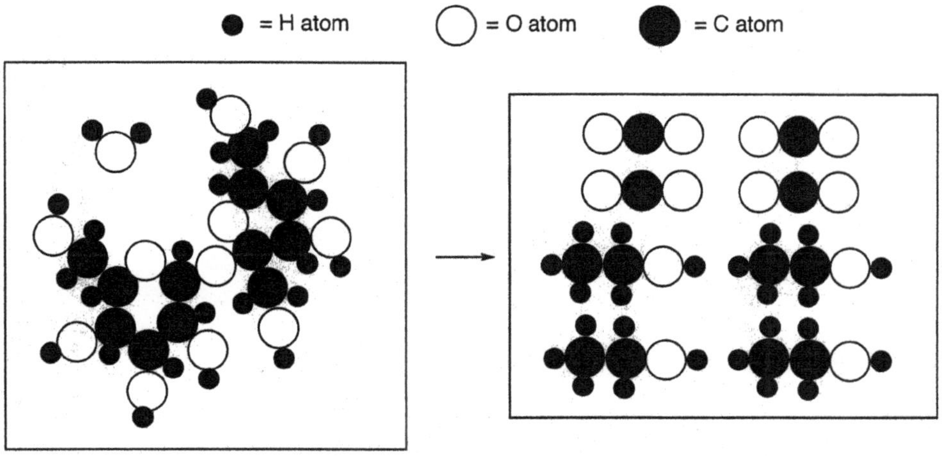

$C_{12}H_{22}O_{11} + H_2O \rightarrow 4\ C_2H_6O + 4\ CO_2$
(sucrose) (water) (ethanol) (carbon dioxide)

10. Most chemists prefer to write this reaction using the formulas instead of the representation with the balls. Why do you think that is?

11. Considering the reaction in model 2, how many total C atoms are there on the:
    a) left side? ______
    b) right side? ______

12. How many total H atoms are there on the:
    a) left side? ______
    b) right side? ______

13. How many total O atoms are there on the:
    a) left side? ______
    b) right side? ______

14. How many total molecules are there on the:
    a) left side? ______
    b) right side? ______

15. Do your answers here confirm or refute your prediction from question 9? Explain.

Model 3: Another chemical reaction:

Burning carbon in the presence of oxygen molecules (O_2) can produce carbon monoxide. Here is a chemical formula representation of that reaction, but it is not balanced:

$$C + O_2 \rightarrow CO$$

Critical Thinking Questions:

16. For each of the reactions written below, decide whether it is a correct or incorrect way to fix the first reaction to make it balanced and circle your answer.

a)	$C + O \rightarrow CO$	Correct	Incorrect
b)	$C + O_2 \rightarrow CO_2$	Correct	Incorrect
c)	$2\,C + O_2 \rightarrow 2\,CO$	Correct	Incorrect
d)	$12\,C + 6\,O_2 \rightarrow 12\,CO$	Correct	Incorrect

Explain why the new reactions are correct or incorrect:

a)

b)

c)

d)

17. For the reaction above, draw a diagram similar to the ball representations shown in Models 1 and 2 to show the balanced chemical transformation.

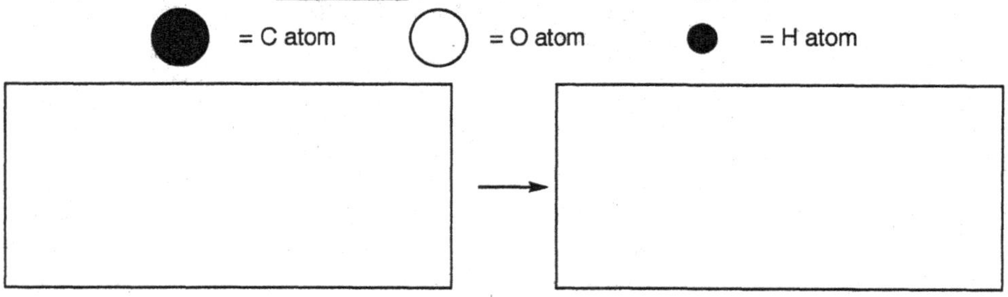

Exercises
(Note: Cl is **chlorine** – *not* carbon and iodine!)

(a) $CH_4 + Cl_2 \rightarrow CH_3Cl + HCl$

(b) $C_3H_8 + O_2 \rightarrow CO_2 + H_2O$

(c) $ZnO + HCl \rightarrow ZnCl_2 + H_2O$

1. For the above reactions, check if they are balanced by counting the number of each type of atom on the left and right sides of the arrow. If they are not balanced, add the correct coefficients and balance them.

2. Here is a list of chemical reactions to balance by adding the correct coefficients. Because there are so many different types of chemical reactions, there is no algorithm or formula or system to follow when balancing reactions: it requires a lot of trial and error, guessing and checking, and experience.

a. ___NO + ___O_2 → ___NO_2

b. ___Al + ___Cl_2 → ___$AlCl_3$

c. ___H_2O_2 → ___H_2O + ___O_2

d. ___S + ___O_2 → ___SO_3

e. ___C_2H_6 + ___O_2 → ___CO_2 + ___H_2O

f. ___NH_3 + ___Cl_2 → ___NH_4Cl + ___NCl_3 (hard!)

g. ___ C_2H_5OH + ___O_2 → ___H_2O + ___CO_2

h. ___ Cu + ___ $FeCl_3$ → ___ $CuCl_2$ + ___Fe

CHEMISTRY 100: ChemActivity 3
Nuclear Reactions

Group Name ________________ Recorder Name ______________

When scientists speak of a *chemical reaction*, they are almost always talking about cases like those in the last ChemActivity, where atoms in various compounds are rearranged, but the atoms themselves are not changed. However, there is another important class of reactions – **nuclear reactions** – where the nuclei themselves are altered, transmuting one element into another.

Model 1: Examples of isotopes seen in nuclear chemistry:

Isotope	Atomic mass (amu)	Number of protons	Number of neutrons
$^{238}_{92}U$	238	92	146
$^{235}_{92}U$	235	92	
	90	38	
	3		2
		28	32
$^{1}_{0}n$			

Note: $^{238}_{92}U$ can be written as "U-238" or "Uranium 238".

Critical Thinking Questions:

1. What does the '92' in $^{238}_{92}U$ represent?

2. What does the '238' in $^{238}_{92}U$ represent?

3. How did we know from the symbol $^{238}_{92}U$ that it contains 146 neutrons?

4. Fill in the missing <u>numbers</u> in the table (atomic mass, number of protons, and number of neutrons), except, for now, for the bottom row.

5. Recall that another name for the number of protons is the **atomic number**. Use that information, plus the values in the table in Model 1 and a periodic table to write the missing elemental symbols in the first column of the table. Write them in the same style as $^{238}_{92}U$.

6. Complete the bottom row of the table. What do you suppose $^{1}_{0}n$ represents?

Model 2: Nuclear reactions: Fission and Fusion

Nuclear **fission** involves the splitting of one larger atomic nucleus into two or more smaller components, typically with a release of a large amount of energy. In many cases fission is initiated by the addition of a neutron.

Nuclear **fusion** involves the joining together of two smaller nuclei into a larger one, again typically releasing a large amount of energy.

Example 1: $^{235}_{92}U + ^{1}_{0}n \rightarrow ^{141}_{56}Ba + ^{92}_{36}Kr + 3\ ^{1}_{0}n$

Example 2: $^{1}_{1}H + ^{2}_{1}H \rightarrow ^{3}_{2}He$

Critical Thinking Questions:

7. Add up the superscripts on the left and right of Example 1.

8. Add up the subscripts on the left and right of Example 1.

9. Why do you think both the total of superscripts and subscripts remain the same during the nuclear reaction?

10. Confirm that the same balanced equation holds for Example 2.

11. Classify examples 1 and 2 as either fission or fusion.

Model 3: Nuclear Reactions: Alpha and Beta decay

An **alpha particle** is another name for a helium-4 nucleus, ${}^{4}_{2}\mathrm{He}$. Many very large nuclei spontaneously decay by emission of alpha particles (called **alpha-decay**).

Example 3: ${}^{239}_{94}\mathrm{Pu} \rightarrow {}^{235}_{92}\mathrm{U} + {}^{4}_{2}\mathrm{He}$

A **beta particle** is an electron (e^-). It can be generated in a nuclear reaction called **beta-decay** in which one of the neutrons in the nucleus turns into a proton $\left({}^{1}_{1}\mathrm{p} \text{ or } {}^{1}_{1}\mathrm{H}\right)$ and releases a high-speed electron, ${}^{0}_{-1}\mathrm{e}$:[1]

Example 4: ${}^{1}_{0}\mathrm{n} \rightarrow {}^{1}_{1}\mathrm{H} + {}^{0}_{-1}\mathrm{e}$

Example 5: ${}^{60}_{27}\mathrm{Co} \rightarrow {}^{60}_{28}\mathrm{Ni} + {}^{0}_{-1}\mathrm{e}$

Critical Thinking Questions:

12. Confirm that all reactions are balanced in terms of atomic masses (superscripts) and atomic number (subscripts)

13. Using the examples above as analogies, what would be the products of the alpha decay of ${}^{238}_{92}\mathrm{U}$? Write a balanced nuclear reaction which shows this.

14. What would be the products of the beta decay of ${}^{131}_{53}\mathrm{I}$? (used in thyroid imaging) Write a balanced nuclear reaction which shows this.

15. It looks like the electron has an atomic number of "-1". Does that make sense? How can it have minus-1 protons?

[1] This treatment of beta decay is not complete: it doesn't include the production of antineutrinos nor describe the involvement of W-bosons or the weak nuclear force in this process. Nevertheless, this treatment of beta-decay is adequate for writing balanced nuclear reactions.

16. Write balanced nuclear reactions for the following processes:

a) The alpha-decay of Americium-241 (this process occurs in common smoke detectors)

b) The beta-decay of Carbon-14 (used indirectly to analyze the age of historical artifacts)

Exercises

1. Write <u>balanced</u> nuclear reactions according to the descriptions below.

   a) After activation by a neutron, U-235 undergoes fission to produce Ce-144, Se-89, and some neutrons.
   b) In the core of the sun, two He-3 nuclei fuse together to product a He-4 nucleus and two protons (a proton can also be called H-1).
   c) Radon-222 decays, producing an alpha particle (a He-4 nucleus) and another nucleus.

CHEMISTRY 100: ChemActivity 4**
Unit conversions

You know how you sometimes hear things like "If everyone in America stood on top of each other's shoulders, you could build a column as high as the moon". How do you think people come up with statements like that? Believe it or not, all that's involved is the use of CONVERSION factors, which are used to either change a measurement from one unit to another, or to relate equivalent items. Sometimes this process is called **unit analysis**.

Model 1:

- Dr. Nichols drives a Honda Accord.
- The car can travel 28 miles per one gallon of gasoline.
- Its fuel tank can hold 16 gallons of gas.
- Currently at the local Shell station, gas costs $2.95 per gallon.

Critical Thinking Questions

1. Three values are given in Model 1. One of them gives a measured **quantity**. Write that quantity here with its associated unit.

2. Two of the statements above give **ratios**. Circle those two statements.

3. Write each ratio that you circled in Model 1 as a fraction. Your fraction needs a number **and** a unit in both the numerator and the denominator. An example of how to write such a fraction (unrelated to Model 1) is shown:

$$\frac{70 \text{ miles}}{1 \text{ hour}}$$

4. How many miles can Dr. Nichols travel on a full tank of gas? Explain how you used the quantity (CTQ 1) and a ratio (CTQ 3) to get your answer. Show all units.

** Adapted from ChemActivity 4 in Garoutte, M. P. *General, Organic, and Biological Chemistry: A Guided Inquiry.* (Wiley, 2007).

5. What happened to the unit "gallons" in your calculation in CTQ 4?

6. We can call fractions like those **conversion factors**. Explain what that means.

7. Here are some conversion factors. Do they all give the same information? Explain why or why not.

$$\frac{28 \text{ miles}}{1 \text{ gallon}} \qquad \frac{28 \text{ mi}}{1 \text{ gal}} \qquad \frac{28 \text{ mi}}{\text{gal}} \qquad \frac{1 \text{ gallon}}{28 \text{ miles}}$$

8. How much money will it cost Dr. Nichols to fill up his tank? Show your work in the same manner as you did in CTQ 4.

Model 2: Feet, inches, and centimeters

1 foot = 30.48 cm (exactly)

3 feet = 1 yard

Critical Thinking Questions

9. Draw a large **X** through each incorrect conversion factor below.

$$\frac{30.48 \text{ cm}}{1 \text{ foot}} \qquad \frac{30.48 \text{ ft}}{1 \text{ cm}} \qquad \frac{1 \text{ cm}}{30.48 \text{ ft}} \qquad \frac{1 \text{ ft}}{30.48 \text{ cm}}$$

10. Explain your answer to CTQ 9.

11. Suppose you wanted to convert a length from feet into centimeters. Which conversion factor from CTQ 9 would you choose? Explain your answer.

Model 3: The Unit Plan

A unit plan begins with a **unit of known quantity** and shows how the units will change after the use of each conversion factor. They can have one single unit conversion or they can be more complicated.

Example 1: Converting feet into centimeters

Unit Plan: ft → cm

Example 2: Let's say you wanted to know how much money it would cost you to buy 6 boxes of tomatoes (which are sold by the pound).

Unit Plan: boxes → pounds → dollars

Critical Thinking Questions

12. The distance from wicket to wicket on a cricket pitch its 22 yards. Since cricket is mostly played in countries which use the metric system, the distance should be reported in centimeters.

    a) Complete the unit plan for converting the length into centimeters:

    yards → ______ → ______

    b) Write as fractions the two conversion factors that correspond to each arrow in the unit plan in CTQ 12(a) – see Model 2 for the values.

    c) Perform the calculation by multiplying the initial **quantity** by each conversion factor in order. Show all work, including units.

13. My summer lab research assistants each worked 10 weeks, working 36 hours per week. Their salary was $9.50 per hour. How much did they each earn during the summer?

    a) Complete a unit plan. Remember to start with the **known quantity**.

    ______ → ______ → ______

    b) Write as fractions the two conversion factors that correspond to each arrow in the unit plan above.

    c) Perform the calculation by multiplying the initial **quantity** by each conversion factor in order. Show all work, including units.

Exercises:

Some of these questions may look more difficult than the ones done previously. For each question, **make a unit plan**, write down each conversion factor, and then do the calculation: if you proceed systematically you should get the correct answer, no matter how many intermediate steps you need. Any conversion factor you need for the problems below are either given or were discussed in lecture on Monday.

1. Convert the distance of 0.022 km to cm.

2. Using the values in Model 1, calculate how much money it will cost for gas for Dr. Nichols to drive (one-way) to his mother-in-law's house, which is 62 miles away. Round off to the nearest penny.

3. If everyone in Canada stood on top of each other's heads, how many miles high would the tower reach? Note that there are 33,000,000 people in Canada. Assume each person is 5 feet tall. Note that a mile is 5280 feet.

4. How many H atoms would you have to string end-to-end to reach a length of one foot? Each H atom is 7.4×10^{-11} m across. Write your unit plan carefully.

5. Barry Bonds hit 762 home runs, which is a career record. What has a greater mass, Barry Bonds (228 lbs) or the combined mass of those 762 baseballs? Info you'll need; each ball weighs 143 grams, and there are 453.6 grams in 1 lb. *Hint: there are two ways to approach this problem. Discuss them with the rest of your group.*

6. How many atoms are found in a pure gold coin that has a mass of 29.0 grams? *The conversion factors for this question were discussed on Monday. Think moles!*

CHEMISTRY 100: ChemActivity 5:
Solutions and Dissolving

GROUP NAME ____________________ RECORDER ______________

Demonstration 1:

Watch as the instructor mixes a sample of sugar (sucrose, formula $C_{12}H_{22}O_{11}$) into a sample of water.

Vocabulary: sucrose is an example of a **non-electrolyte**.

Critical Thinking Questions:

1. What happened on a macroscopic level (*i.e.* what did you observe)

2. Which of these diagrams most accurately represents the final solution on a microscopic level?
Note: the formula for water is H_2O. The formula for sucrose is $C_{12}H_{22}O_{11}$

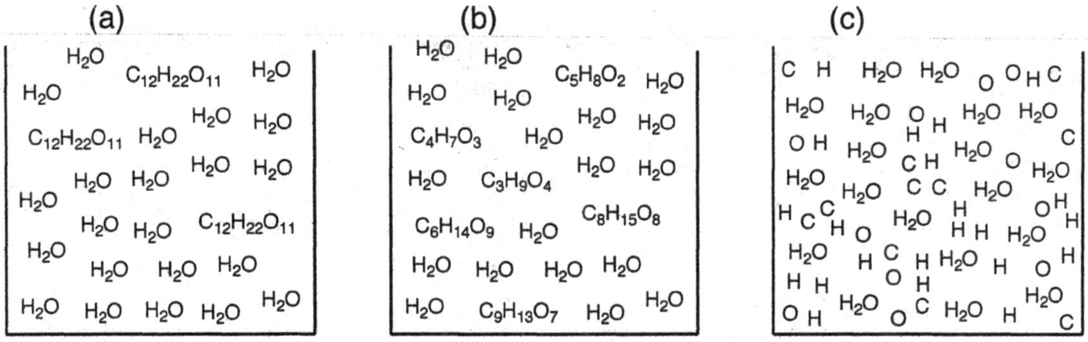

3. Explain your answer to Question 2.

4. Is there a simple test you could do to verify your prediction? If enough groups come up with a simple enough test the instructor will do it...

Demonstration 2:

Watch as the instructor mixes a sample of salt (NaCl) into water.

Vocabulary: Sodium chloride is an example of an **electrolyte.**

Critical Thinking Questions:

5. What happened on a macroscopic level (*i.e.* what did you observe)

6. Which of these diagrams most accurately represents the final solution on a microscopic level?
Note: the formula for water is H_2O. The formula for salt is NaCl.

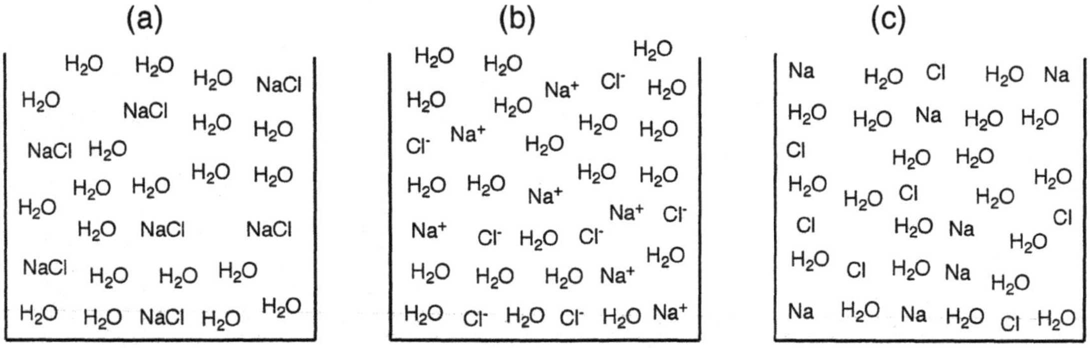

7. Explain your answer to Question 6.

8. Is there a simple test you could do to verify your prediction? If enough groups come up with a simple enough test the instructor will do it...

9. Why are there such different answers for these two very similar situations?

10. Summarize the instructor's comments on the difference between these two situations.

11. For each of the compounds listed below, predict whether they would result in a situation like that for sugar or like that for salt.

a) ethanol (C_2H_6O) __________
b) calcium chloride ($CaCl_2$) __________
c) iodine (I_2) __________
d) sodium hydroxide (NaOH) __________ (this one is tricky)

Why is sodium hydroxide tricky? ______________________________

Demonstration 3:

The instructor will place both containers of sugar solution and salt solution on hot plates that have a temperature over 100 °C. You should be able to see some mist rising off the plates:

Critical Thinking Questions:

12. What do you think will happen with the sugar solution? (Multiple choice)
(a) The water will evaporate, taking the sugar with it. The container should become empty.
(b) The sugar will prevent the water from evaporating. Nothing will happen.
(c) The water will boil away, leaving a solid sugar deposit at the bottom of the container.
(d) Something else

13. After a while, report what actually did happen to the mixture. Explain why you think it happened how it did.

14. What do you think will happen with the salt solution? (Multiple choice)
(a) The water will evaporate, taking the salt with it. The container should become empty.
(b) The salt will prevent the water from evaporating. Nothing will happen.
(c) The water will boil away, leaving a solid salt deposit at the bottom of the container.
(d) Something else

15. After a while, report what actually did happen to the mixture. Explain why you think it happened how it did.

Demonstration 4:

The instructor will mix two solutions together. One contains a solution of silver nitrate ($AgNO_3$). The other contains a solution of sodium chloride (NaCl).

Critical Thinking Questions:

Before the demo:

16. Are these solutions composed of ions or molecules? *Note: "nitrate" (NO_3^-) is an example of a* ***polyatomic ion****, which consists of a number of non-metal atoms bound together which have an overall charge.*

17. Draw diagrams of what these solutions look like on a microscopic scale. You can omit the water molecules in the solution for the sake of clarity. Also predict what the solution will look like when the samples are mixed.

$AgNO_3$ (aq.) NaCl (aq.) Combined

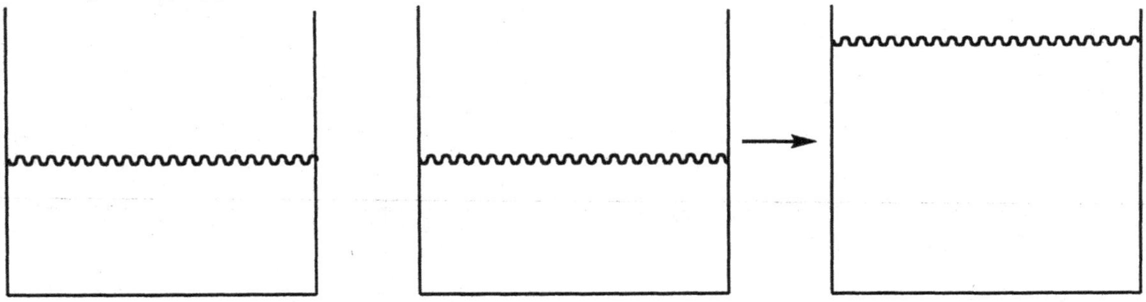

After the demo:

18. What did you observe?

19. Which of the diagrams below do you think best explains what happened? (the lump on the bottom represents the solid that is formed)

(a) (b) (c) (d)

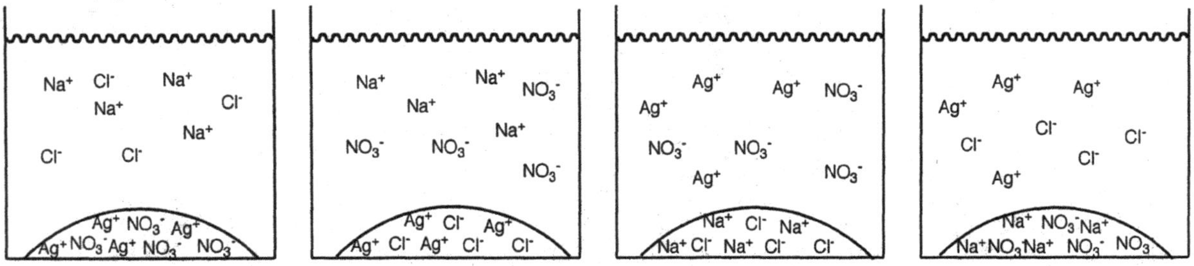

20. How could you verify your answer to question 19 experimentally?

Model 1:

Imagine that what is shown below is 1 liter of a sugar solution (S = sucrose; water molecules omitted for the sake of clarity).

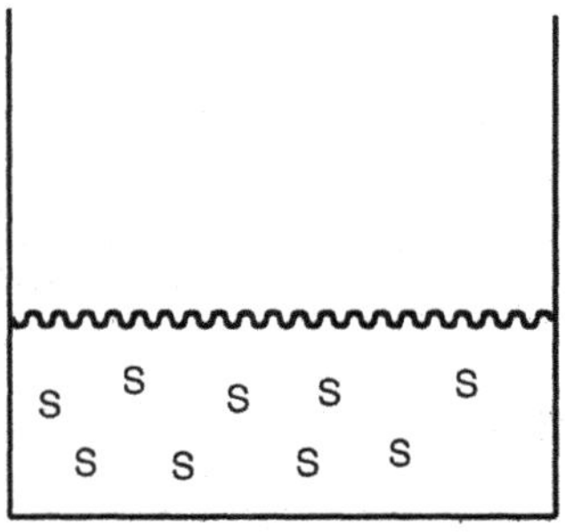

Critical Thinking Question:

21. What would the resulting solution look like if an additional 1 liter of water was added to this solution?

(a) (b) (c)

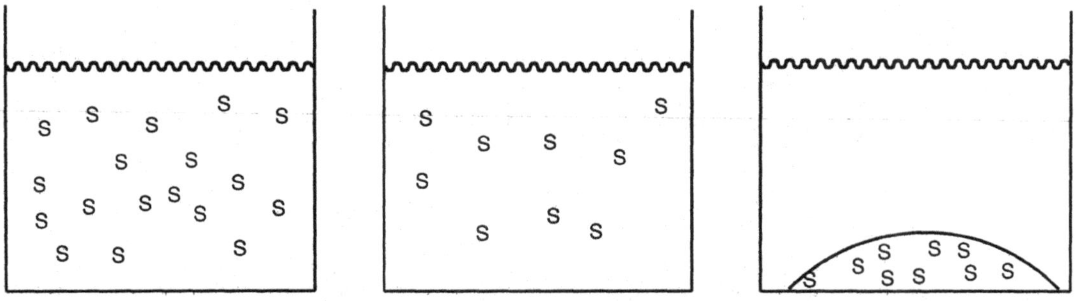

22. Has the solution become more concentrated? more dilute? neither?

23. Explain your answer to question 22.

24. If the initial concentration of the 1 liter sugar solution was 900 ppm, what would be the final concentration of the solution in question 21? (No calculators should be needed). Explain your answer.

CHEMISTRY 100: ChemActivity 6:
Acids and Bases

Model 1: GROUP NAME ____________ RECORDER ____________

Below are diagrams of three different aqueous solutions.

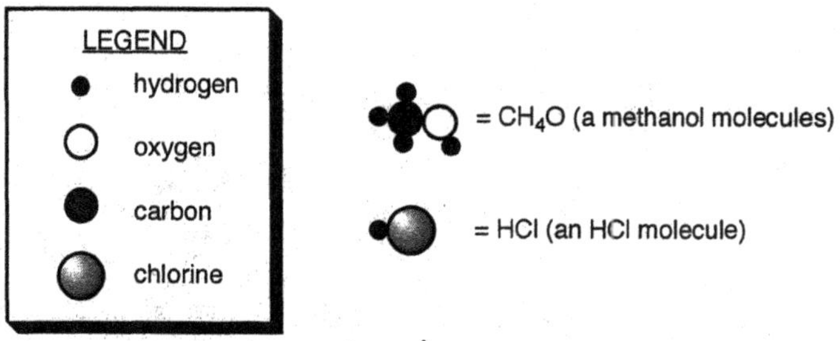

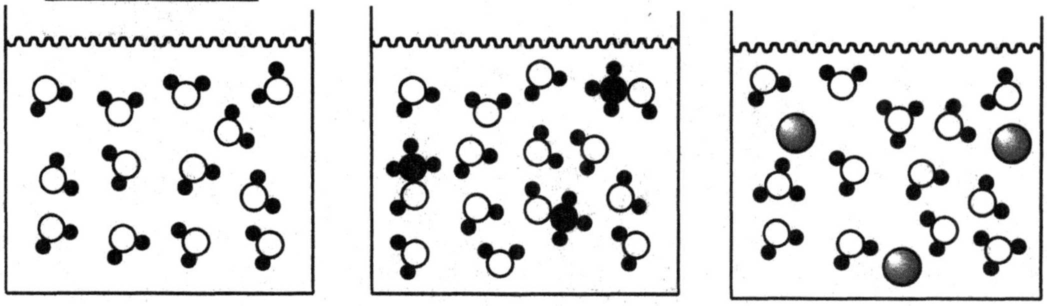

Water (pure) | Methanol (CH_4O) dissolved in water | HCl dissolved in water

Critical Thinking Questions

1. What happened to the methanol molecules when they were dissolved in water?

2. What happened to the HCl molecules when they dissolved in water?

3. Suppose I told you that the HCl solution was a very good conductor of electricity. What would that tell you about the solution?

4. What molecules or ions do these figures represent?

 =

 =

5. Complete the reaction of HCl with H_2O:

$HCl_{(g)} + H_2O_{(l)} \longrightarrow$

Definitions:

An **acid** is any substance that causes an increase in the concentration of H_3O^+ ions when dissolved in water.

A **base** is any substance that causes an increase in the concentration of OH^- ions when dissolved in water.

Critical Thinking Questions

6. Is methanol an acid, a base, or neither? Explain your answer.

7. Is HCl an acid, a base, or neither? Explain your answer.

8. What molecules/ions are present in solution when sodium hydroxide (NaOH) is dissolved in water? Use that answer to classify NaOH as an acid, base, or neither.

Model 2: pH

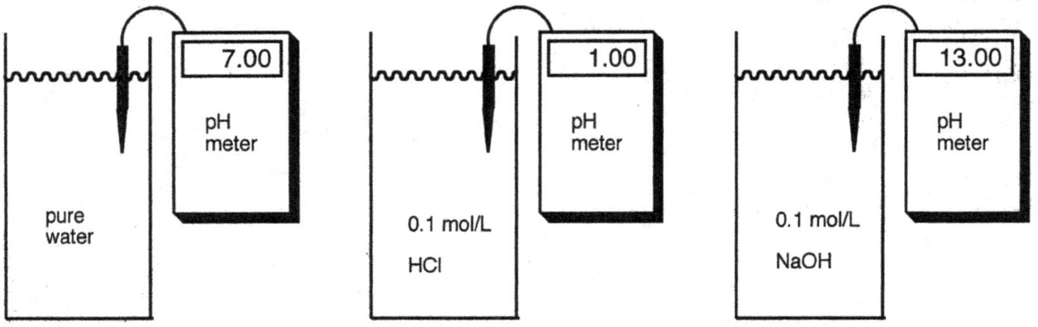

The pH scale is a quantitative measure of the acidity or basicity of an aqueous solution. A pH of 7.00 is considered neutral (i.e. neither acidic nor basic).

Critical Thinking Questions

Circle the correct answer.

9. The pH of an acid solution is (higher lower) than the pH of pure water.

10. The pH of a base solution is (higher lower) than the pH of pure water.

Model 3: pH vs. molarity of H_3O^+

Remember "molarity" just means concentration expressed in $\frac{mol}{L}$.

Concentration of H_3O^+ ($\frac{mol}{L}$)	Concentration of H_3O^+ in scientific notation	pH	Acidic, Neutral, or Basic?
1.0	1.0×10^{0}	0	
0.1	1.0×10^{-1}	1	
0.001		3	
	1.0×10^{-5}	5	
0.0000001			
		11	
		14	

Critical Thinking Questions

11. Based on the entries in the first two lines of the table, comment on the numerical relationship between the molarity of H_3O^+ ions in solution and pH.

12. Complete the table by filling in all the other blanks.

13. The last two entries have only very small concentrations of H_3O^+ ions. What ions do they have greater concentrations of? Explain your answer.

14. As the concentration of H_3O^+ increases, the pH ____________.

15. If the concentration of H_3O^+ goes up by a factor of 10, then the pH goes __________ by ________ units.

16. If the concentration of H_3O^+ goes down by a factor of 100, then the pH goes __________ by _________ units.

17. Even in a neutral solution, there is still a small concentration of H_3O^+ ions. What is the concentration of H_3O^+ in neutral water?

Exercises:

Please read **Section 8.2** in the online textbook http://preparatorychemistry.com/Bishop_Atoms_First.htm for much more information on pH.

1. Acid rain is caused by both sulfuric acid (H_2SO_4) and nitric acid (HNO_3) being formed in the air and dissolving in atmospheric water before coming down to the surface as rain.

   a) Write a reaction of nitric acid with water to illustrate its acidity.

   b) Write a reaction of sulfuric acid with water to illustrate its acidity.

2. The pH of rainwater near the great lakes is the most acidic in the USA. The pH can get as low as 4.1. Normal rainwater has a pH of about 5.3.

   a) Calculate the approximate concentration of H_3O^+ of acid rain.

   b) Calculate the approximate concentration of H_3O^+ of normal rain.

   c) In the pH scale, pure water has a pH of 7.0. Even normal rainwater has a lower pH than 7.0 (typically it's around 5.6). What is the source of the acidity even in rainwater that has no sulfuric acid or nitric acid in it?

3. H_3O^+ ions and OH^- ions cannot both be present in large concentrations in aqueous solution: a reaction occurs between them, NEUTRALIZING the acid and the base. Complete the reaction showing the neutralization of H_3O^+ with OH^-.

$$H_3O^+{}_{(aq)} + OH^-{}_{(aq)} \longrightarrow$$

CHEMISTRY 100: ChemActivity 7
Science and the Scientific Method

GROUP NAME ______________ RECORDER ______________

In today's activity there are no 'models' to gather information from. Today we will be discussing people's perception of science, the scientific method, and scientists. After the groups have finished discussing the various questions amongst themselves, we will ask the spokespersons to talk about their answers, and see what people agree and disagree on. There are no 'right answers' in this activity.

Critical Thinking Questions.

For each statement below, your group should work together to decide if they think the statement is **True or False** (either by consensus or by majority opinion). Each answer must be accompanied by an explanation and/or examples.

1. There are certain areas of study that should be off-limits to science and scientists.

2. Scientists are all out to push a particular agenda like everyone else.

3. So many scientific theories have changed over time that there's no reason to trust today's scientists with their current theories, since tomorrow they could all be proven wrong.

4. Since we weren't around to make observations 1,000,000 years ago, you can't really say what happened back then.

5. Scientists are stuck in their ways: they'll automatically reject any new sounding hypothesis that comes along.

6. In order for students to effectively evaluate a scientific claim, they should hear "both sides of the story" (from people are agree and those who disagree with the claim).

7. Science can only be done by people with plenty of scientific education.

8. Science is our best method of acquiring knowledge.

9. Extraordinary claims require extraordinary evidence.

10. A good way to attack critics of science is by making fun of them.

11. A good way to prop up an argument is to cite authorities and experts.

12. "Nobody has proven it wrong" is a sensible reason to believe something is right.

13. Science is a religion.

14. Science is responsible for nuclear bombs and trans fats and many other things harmful to humanity – we need to get away from science and simplify our lives to live longer and happier

15. Science should be an important part of the curriculum for children in elementary school.

CHEMISTRY 100: ChemActivity 8
Baloney Detection and Logical Fallacies

Model 1: The Baloney Detection Kit

Noted science popularized Carl Sagan, author of such books as *Cosmos* and *Contact*, has devised a set of tools to help people distinguish reasonable, scientific ideas from non-scientific ones. A synopsis of this part of his book (*The Demon Haunted World*) is presented here. Please read this passage:

1. Wherever possible there must be independent confirmation of the "facts."
2. Encourage substantive debate on the evidence by knowledgeable proponents of all points of view.
3. Arguments from authority carry little weight -- "authorities" have made mistakes in the past. They will do so again in the future. Perhaps a better way to say it is that in science there are no authorities; at most, there are experts.
4. Spin more than one hypothesis. If there's something to be explained, think of all the different ways in which it could be explained. Then think of tests by which you might systematically disprove each of the alternatives. What survives, the hypothesis that resists disproof in this Darwinian selection among "multiple working hypotheses," has a much better chance of being the right answer than if you had simply run with the first idea that caught your fancy.
5. Try not to get overly attached to a hypothesis just because it's yours. It's only a way station in the pursuit of knowledge. Ask yourself why you like the idea. Compare it fairly with the alternatives. See if you can find reasons for rejecting it. If you don't, others will.
6. Quantify. If whatever it is you're explaining has some measure, some numerical quantity attached to it, you'll be much better able to discriminate among competing hypotheses. What is vague and qualitative is open to many explanations. Of course there are truths to be sought in the many qualitative issues we are obliged to confront, but finding them is more challenging.
7. If there's a chain of argument, every link in the chain must work (including the premise) -- not just most of them.
8. Occam's Razor. This convenient rule-of-thumb urges us when faced with two hypotheses that explain the data equally well to choose the simpler.
9. Always ask whether the hypothesis can be, at least in principle, falsified. Propositions that are untestable, unfalsifiable are not worth much.

Group Name ____________________ Recorder Name __________________

Critical Thinking Questions

1. In each of the cases below, an unscientific or illogical argument is being used. Use the tools of the baloney detection kit to decide how to discredit the argument. *Note: the statements being made may in fact be true, but the arguments used to arrive at the conclusions are not logically sound. Don't attack the statement: focus on the methods used to arrive at the conclusion.*

a) There is another dimension out there inhabited by spirits. I know it is there because I can 'feel' it, even though it is not detectable by my normal senses or any of our modern scientific instruments.

b) (Argued by a juror in a trial) I have decided that the defendant is guilty. The prosecutor's opening statement had me convinced right away. I heard some counter arguments from the defense attorney but rejected them because there really was no way to explain what happened other than what the prosecutor said at the start of the trial.

c) We should blame advertisers from the obesity epidemic that is occurring in the USA. After all, if they weren't constantly pounding on us to eat their junk food, we wouldn't eat it, so we wouldn't all gain so much weight.

d) Our Universe and everything in it is just one of an infinite number of other universes in existence. The Big Bang theory says that the universe is finite, but since all that exists is probably infinite, there is lots of room for other universes, so it seems very likely that they are there.

Model 2: Logical Fallacies

Below is a partial list of fallacies used when arguing a claim illogically or unscientifically, and an example or two is given for each one. Note that there are many other logical fallacies (which you may have encountered in a course in philosophy) – this list is not meant to be complete. Please read through it.

Ad hominem
Latin for "to the man", attacking the arguer and not the argument.
Example: *The Reverend Dr. Smith is a known Biblical fundamentalist, so her objections to evolution need not be taken seriously.*

Argument from authority
Determining the truth of a statement based on who made the statement.
Example: *Dr. Phil says "Anger is nothing more than an outward expression of hurt, fear and frustration", so it must be true.* Authorities can be wrong, even when discussing ideas relevant to their field of expertise.

Appeal to ignorance
The claim that whatever has not been proved false must be true, and vice versa.
Example: *There is no compelling evidence that UFOs are not visiting the Earth; therefore UFOs exist* .

Special pleading
Essentially, this involves someone attempting to cite something as an exemption to a generally accepted rule, principle, etc. without justifying the exemption. Often used to rescue a proposition in deep rhetorical trouble.
Example: *Meditation can't make you levitate: that defies the laws of physics.* Special plead: *When you're in your trance, the laws of physics don't apply.*

Begging the question
Also called assuming the answer.
Example: *We must institute the death penalty to discourage violent crime.* This assumes (without proof) that violent crime is discouraged by the death penalty: this is 'circular reasoning'.

Observational selection
Also called the enumeration of favorable circumstances, or as the philosopher Francis Bacon described it, counting the hits and forgetting the misses.

Statistics of small numbers
A close relative of observational selection. E.g., *I've thrown three sevens in a row. Tonight I can't lose.*

Non Sequitur.
Latin for "It doesn't follow". The premise and the conclusion aren't logically related. Example: *Our jihad will prevail because Allah is great.* Often those falling into the non sequitur fallacy have simply failed to recognize alternative possibilities.

Excluded middle, or False Dichotomy
Considering only the two extremes in a continuum of intermediate possibilities. Example: *Either you support the president's decision to go to war or you are with the terrorists.*

Confusion of correlation and causation
Assuming that because two things go together that one caused the other.
Example: *A survey shows that more college graduates are homosexual than those with lesser education; therefore education makes people gay.*
Example #2: *The bigger a child's shoe size, the better the child's handwriting.* Does having big feet make it easier to write? No, it means the child is older.

Straw man
Caricaturing or over-simplifying a position to make it easier to attack.
Example: *Environmentalists care more for snail darters and spotted owls than they do for people.*

Critical Thinking Questions:[**]
Please note that the objective here is not for you to memorize the specific names of these logical fallacies – it is for you to learn to recognize these illogical arguments when you encounter them, and to do that it helps to compare new arguments with others you have already seen.

2. Each of the statements below uses one or more of the logical fallacies mentioned on the previous page. Decide which fallacy it is, and offer a one-sentence counter-argument to refute the statement.

a) Quantum mechanics doesn't make any sense: even Einstein thought it was rubbish.

b) When sales of hot chocolate go up, street crime drops. Therefore hot chocolate prevents crime.

** Many of these examples are taken from this website: http://www.don-lindsay-archive.org/

c) I've lost my keys AGAIN. It seems like every day I lose my keys!

d) You either love organic chemistry or you hate it.

e) Of course I do not support amnesty for illegal immigrants. George Bush supports amnesty, and he's an idiot.

f) If you have Budweiser at your next party, everyone will have a good time!

g) The supporters of medical marijuana legislation are really just a bunch of stoners looking to legalize their lazy lifestyle.

h) Cosmologists have shown that 96% of the universe is "dark matter" and "dark energy", which they do not yet understand. Mainstream scientists are not presently looking for "dark matter" in man. They need to. This could finally provide a scientific link between the mind, body, and soul. (see http://www.esolibris.com/articles/science_spirituality/dark_matter_01.php)

i) Murder is morally wrong. Therefore, abortion is morally wrong.

j) Dr. Nichols and Mrs. Nichols have a policy whereby if one cooks dinner, the other does the dishes:
Mrs. Nichols: "I cooked tonight, so it's your turn to do the dishes"
Dr. Nichols: "The baseball playoffs are on, so I shouldn't have to do the dishes"

k) This tea tree oil is far better than aspirin in getting rid of headaches. My friend told me about how well it worked for her, so I tried it one day when I had a nasty headache, and it went away!

3. See if you can write some illogical statements that might sound logical to someone unfamiliar with the fallacies being discussed. Have your examples correspond to the fallacies listed here:

a) Ad hominem

b) Appeal to ignorance

c) Begging the Question

d) Non sequitur

CHEMISTRY 100: ChemActivity 9
Covalent bonding and Lewis Structures

GROUP NAME ________________ RECORDER ______________

Today we'll be looking at covalent compounds and how to predict how atoms will bond together based on the number of electrons they possess and the organization of those electrons within the atom.

Model 1:

Quantum mechanics shows that electrons in atoms exist in different **energy levels.** Furthermore, for elements whose atomic numbers are 20 or below, the energy levels can accommodate:

Level 1:	Max 2 electrons
Level 2:	Max 8 electrons
Level 3:	Max 18 electrons (however, once it has 8 electrons, the next two get added to Level 4)
Level 4:	Max 32 electrons

The **ground state** electronic configuration of an atom has the electrons in the lowest available energy levels.

Critical Thinking Questions

1. How do you know what total number of electrons an atom (with no electrical charge) will have?

2. In the truncated diagram of the periodic table (Figure 1), write in the number of electrons in each energy level for every element up to and including **Ca**. A few have been done for you as examples

Figure 1: Periodic table with electronic configurations

H 1/0/0								He
Li 2/1/0	Be 2/2/0	2/6/0 means: 2 electrons in Lvl 1 6 electrons in Lvl 2 No electrons in Lvl 3	B	C	N	O 2/6/0	F	Ne
Na	Mg		Al	Si	P	S	Cl	Ar 2/8/8
K	Ca	(Transition metals)						

Definition: The electrons in an atom's highest energy level are its **valence electrons**. They are the ones available for forming bonds. The other electrons in the atom are called **inner-shell** electrons.

3. In the partial periodic table below (Figure 2), write in the number of valence electrons in each atom up to and including calcium. *Don't do Ga through Kr just yet.*

Figure 2: Periodic table with valence electron counts

H								He
Li	Be		B	C	N	O	F	Ne
Na	Mg		Al	Si	P	S	Cl	Ar
K	Ca	(Transition metals)	Ga	Ge	As	Se	Br	Kr

4. Summarize any trends you see in the periodic table with regards to the number of valence electrons in an atom.

5. There is one exception to the general trend you may have seen in CTQ 4. What is it?

6. Using your answers to questions 4 and 5, write in the number of valence electrons in the elements from Ga to Kr (bottom row in Figure 2).

Model 2: Another representation of electronic structure

Here is a model of the electronic configuration of an oxygen atom and an aluminum atom.

Figure 3: Concentric-ring diagram of electronic configuration

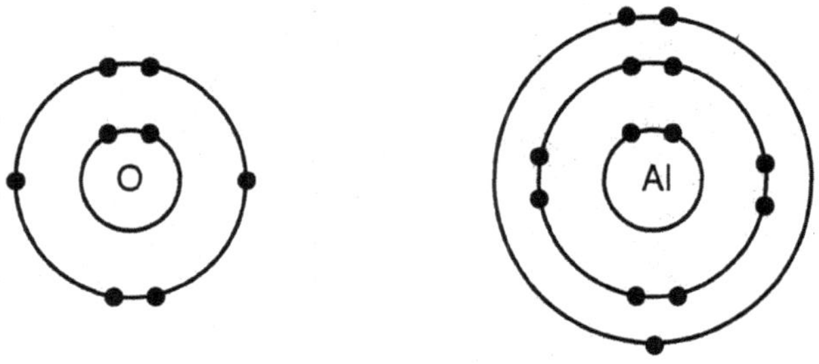

Critical Thinking Questions:

7. What does the letter in the center of the diagram (Figure 3) represent?

8. How are the electrons in level 1 represented?

9. How are the electrons in other energy levels represented?

10. List anything you notice about the way electrons in the same energy level (on the same ring) are drawn.

11. Draw similar diagrams for neon (Ne) and potassium (K). Use Figure 1 to help you.

Model 3: Yet another representation of electronic configuration in atoms

Shown here is another representation of an oxygen atom and an aluminum atom.

Figure 4: Lewis-dot depiction of electronic configuration

Critical Thinking Questions:

12. What does the letter in the center of the diagram represent?

13. How are inner-shell electrons represented (compare Figure 4 with Figure 3)

14. How are valence-shell electrons represented?

15. What makes the "Lewis-dot" representation better than that shown in Figure 3? What makes it worse?

16. Draw Lewis-dot diagrams for all the elements in the periodic table shown. Oxygen and aluminum are done already.

(Use your answer to CTQ 3 (Figure 2) to help you)

Figure 5: Periodic table with Lewis dot atomic structure

H								He
Li	Be		B	C	N	·Ö· (with dots)	F	Ne
Na	Mg		Äl (with dots)	Si	P	S	Cl	Ar
K	Ca	(Transition metals)	Ga	Ge	As	Se	Br	Kr

Model 4: The Octet Rule

Below are some Lewis-dot diagrams of molecules (made of more than 1 atom)

Figure 6: Combining atoms to make molecules

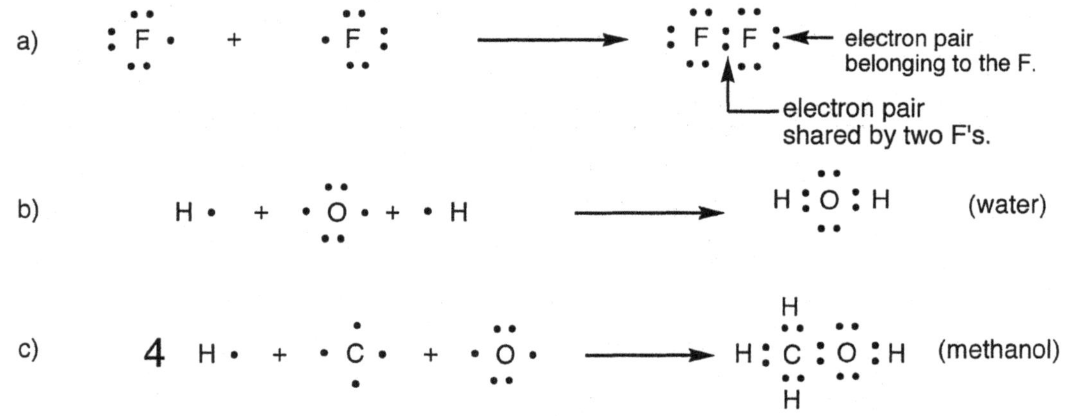

Critical Thinking Questions:

17. For reactions a, b, and c, count up the <u>total</u> number of valence electrons shown in the original atoms (left) and the final molecule (right)

	Valence e^- in ATOMS	Valence e^- in MOLECULES
a		
b		
c		

What can you say about the numbers of valence electrons? ________________

18. In Figure 6, place one circle around each atomic symbol and all of the electrons immediately adjacent to it. In some cases a pair of electrons may be encompassed by more than one circle. An example is shown here:

19. In Figure 6, how many electrons are in the circles for each of these elements?
    a. Hydrogen atom? _______
    b. Carbon atom? ________
    c. Oxygen atom? ________
    d. Fluorine atom? _________

Definition: The **octet rule** says that, for maximum stability, an atom must have 8 valence electrons associated with it. This can include electrons which it shares with other atoms. Hydrogen, helium, and lithium are exceptions to the octet rule: they reach maximum stability with only 2 associated valence electrons.

Definition: A pair of electrons that is associated with only one atom is a **lone pair**. A pair that is shared by two atoms is a **bonding pair**. Bonding pair electrons can be drawn as a line connecting the two atoms rather than as a pair of dots between them. Bonding pair electrons constitute what we call **covalent bonds**.

20. Redraw the Lewis-dot structures of water and methanol (Figure 6), drawing the bonding pairs as a line rather than a pair of dots. F_2 has been done as an example.

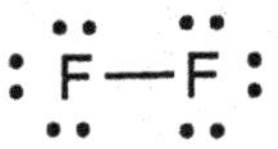

21. Based on the octet rule, why do you think that He, Ne, Ar, and Kr are typically found as lone atoms? In other words, why do they typically not form bonds to other atoms?

Model 5: Lewis Structure of Molecules

Now that you have seen some molecules with all of their valence electrons shown, you will be asked to draw the structure of a variety of molecules. Here are some rules that must be followed when drawing these structures:

A. The total number of electrons shown (including bonding pairs and lone pairs) must be the same as the total number of valence electrons in the atoms that make up the molecule (see CTQ 17 for evidence)
B. In the molecule's structure, all atoms must have EIGHT electrons right next to it: no more, no less. Remember a bond is made up of 2 electrons. The one exception which you will encounter is the hydrogen atom (H), which only needs TWO electrons right next to it.
C. In some cases, the only way to satisfy both of the above rules is to have double or triple bonds (examples to follow)

Examples of good Lewis Structures:

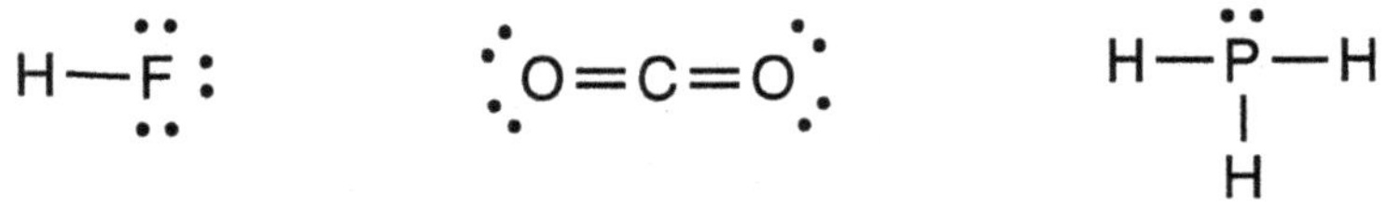

Examples of bad Lewis Structures:

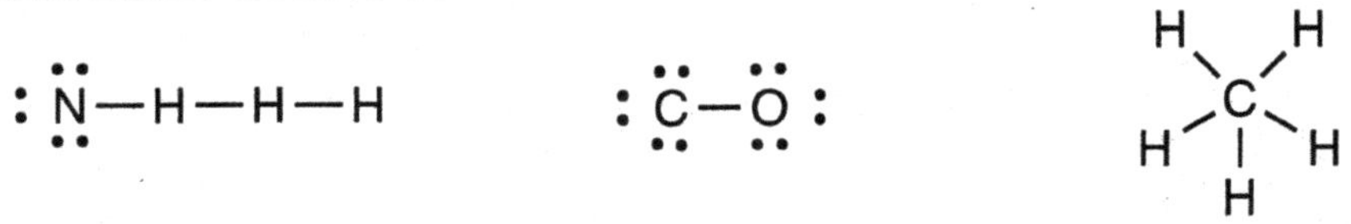

Critical Thinking Questions:

22. For each of the three 'bad' Lewis structures, list two way in which they are not following the rules:

	NH_3	CO	CH_5
1.			
2.		(I think there's only one problem with this one)	

23. For each of the three 'bad' Lewis structures, redraw them (with the molecular formulas shown in CTQ 22) so that they follow all three of the rules listed above under Model 5. *Note: one of them is un-fixable. Can you decide which?*

24. Explain how you can tell that the 'good' Lewis Structure of CO_2 follows both rules A and B shown in Model 5.

Exercises – THIS PAGE TO BE DONE AS HOMEWORK BY EVERYONE

1. Draw Lewis-dot structures for the molecules shown. Keep in mind the rules in Model 5: your molecule can't have more or less valence electrons than the total of the individual atoms which made the molecule, and "stable" atoms have 8 electrons associated with them – except for H (with 2).

a. NH_3

b. $SiCl_4$

c. Br_2

d. C_2H_6

e. CS_2

f. CH_2O

2. Please read the online textbook http://preparatorychemistry.com/Bishop_Atoms_First.htm (sections 5.4 and 5.5), for more information about Lewis structure and covalently bonded molecules.

CHEMISTRY 100: ChemActivity 10
Behavior of Gases

GROUP NAME ______________ RECORDER ______________

Today we will investigate some of the properties of gases, and see how changing one of them (temperature, pressure, volume, etc) changes the others. First, however, we must look at some of the units involved in measuring those quantities.

We will be using the website http://chemconnections.org/Java/molecules/ to model some of the gas behavior. Feel free to link to that and play with it all you want.

Part 1A: Temperature

Critical Thinking Questions

1. If someone asked you "what is *temperature*?", what would you tell them?

2. The instructor will give you the official definition of temperature. Write it down.

3. How does the official technical definition correspond to your initial answer in CTQ1?

Model 1: Temperatures and different units

	°F (Fahrenheit)	°C (Celsius)	K (Kelvin)
Surface of Mars	-67 °F	-55 °C	218.15 K
Ice melts	32 °F	0 °C	273.15 K
Water boils	212 °F	100 °C	373.15 K
Cook a pizza from Papa Murphy's	425 °F	218.33 °C	496.48 K

Conversion factors: $°C = (°F - 32) \div 1.8$ $K = 273.15 + °C$

Critical Thinking Questions

4. How are the conversion factors that change one temperature unit into another different from the conversion factors that we have seen so far (such as grams/mol or miles per gallon)?

5. Can you have a negative number of "miles per gallon"? Can you have a negative number of "grams per mole"?

6. How it is possible to have negative temperature values?

7. Look at the example on the screen of the gas molecules bouncing around.
   a) What happens as you approach zero degrees Kelvin?

   b) What would you see at zero degrees Kelvin?

   c) Why do you suppose that's called "absolute zero"?

Part 1B: Pressure

Critical Thinking Questions

8. If someone asked you "what does it mean to talk about a gas's *pressure*?", how would you answer?

9. By definition, pressure = force per unit area. When talking about gases, what is the source of the force? What "area" is the force acting on? Look at the online molecule page for a clue.

Part 2: How gas pressure changes when you change other variables

For all of these questions, the instructor will set the online bouncing molecules demo to certain settings. Use the pressure gauge at the bottom of the page to answer the questions.

Critical Thinking Questions

10. Set Red to T = 273, Mass = 20 (neon), Number = 80
Set Blue to T = 273, Mass = 131 (xenon), Number = 80

a) What variables have been kept constant between the red and blue gases?

b) What variable has changed? ___________

c) What effect does this change have on the gas pressure?

d) Why do you think the pressure changed the way it did (remember what pressure is, see CTQ 9)?

11. Set Red to T = 273, Mass = 40 (argon), Number = 30
Set Blue to T = 546, Mass = 40 (argon), Number = 30

a) What variables have been kept constant between the red and blue gases?

b) What variable has changed? ___________

c) What effect does this change have on the gas pressure?

d) Why do you think the pressure changed the way it did (remember what pressure is, see CTQ 9)?

12. Set Red to T = 500, Mass = 40 (argon), Number = 20
Set Blue to T = 500, Mass = 40 (argon), Number = 40

a) What variables have been kept constant between the red and blue gases?

b) What variable has changed? ____________

c) What effect does this change have on the gas pressure?

d) Why do you think the pressure changed the way it did (remember what pressure is, see CTQ 9)?

13. Watch the video clip from YouTube: www.youtube.com/watch?v=DcnuQoEy6wA
Initial Pressure in balloon = 1.1 atm. Final Pressure in balloon = 0.3 atm

a) What variables were kept constant during the experiment?

b) What has changed? (besides pressure) ____________

c) What effect did this change have on the gas pressure?

d) Why do you think the pressure changed the way it did (remember what pressure is, see CTQ 9)?

14. Complete the following sentences, using your information from CTQ's 10-13.

Possible answers: *increases* *decreases* *stays the same*

a) As the size of the gas molecules increases, the pressure _______________.

b) As the temperature increases, the pressure _______________.

c) As the number of moles of gas increases, the pressure _______________,

d) As the volume increases, the pressure _______________.

15. The symbol $\propto$ means "is proportional to".
P = pressure; V = volume; T = temperature (in Kelvins); n = "number of moles".
For each pair, circle the answer which corresponded to your results from CTQ's 11 through 13.

a) $P \propto T$ $\quad$ $P \propto \frac{1}{T}$

b) $P \propto n$ $\quad$ $P \propto \frac{1}{n}$

c) $P \propto V$ $\quad$ $P \propto \frac{1}{V}$

16. Combine the three relationships from CTQ 15 into one large expression (all multiplied together):

$$P \propto \_\_\_\_\_ \times \_\_\_\_\_ \times \_\_\_\_\_$$

17. Instead of saying "is proportional to" you can say "equals it multiplied by a constant". Let's call that constant "R". Rewrite your answer to CTQ 16 using R.

$$P = R \times \_\_\_\_\_ \times \_\_\_\_\_ \times \_\_\_\_\_$$

18. Finally, rewrite the equation from CTQ 17 in a condensed algebraic form (without the 'times' symbols)

19. What is the name of the above equation? (If you don't know, the instructor will tell you.)

Exercises

1. If a sample of a gas has P = 1 atm and V = 10 L, and I compress it to V = 2 L, what is the new pressure? Explain your answer.

2. If a sample of a gas has P = 2 atm, and n = 10 mol, and I add more gas until the n = 30 mol, what is the new pressure? Explain your answer.

3. Find the flaw in this argument: "You said that if you double the temperature, you'd double the pressure of an ideal gas. I can prove you wrong: let's say you start at 0 °C, doubling the temperature still gives you 0 °C (since 0 x 2 = 0). The temperature hasn't changed at all so of course the pressure can't have changed either!"

CHEMISTRY 100: ChemActivity 11: Heat transfer

Group Name _______________ Recorder _______________

Today we will be looking more closely at what can happen to a pure substance when heat is added.

Model 1: Heating curve for cyclohexane

Cyclohexane has a molecular formula C_6H_{12}. Below is shown a diagram of what happens to the temperature of a sample of cyclohexane as heat is added at a constant rate by an external source of heat. The same number of Joules are being added to the sample of cyclohexane each minute as the experiment progresses.

The cyclohexane starts out as a solid at -10 ˚C.

Figure 1: Heating curve of cyclohexane

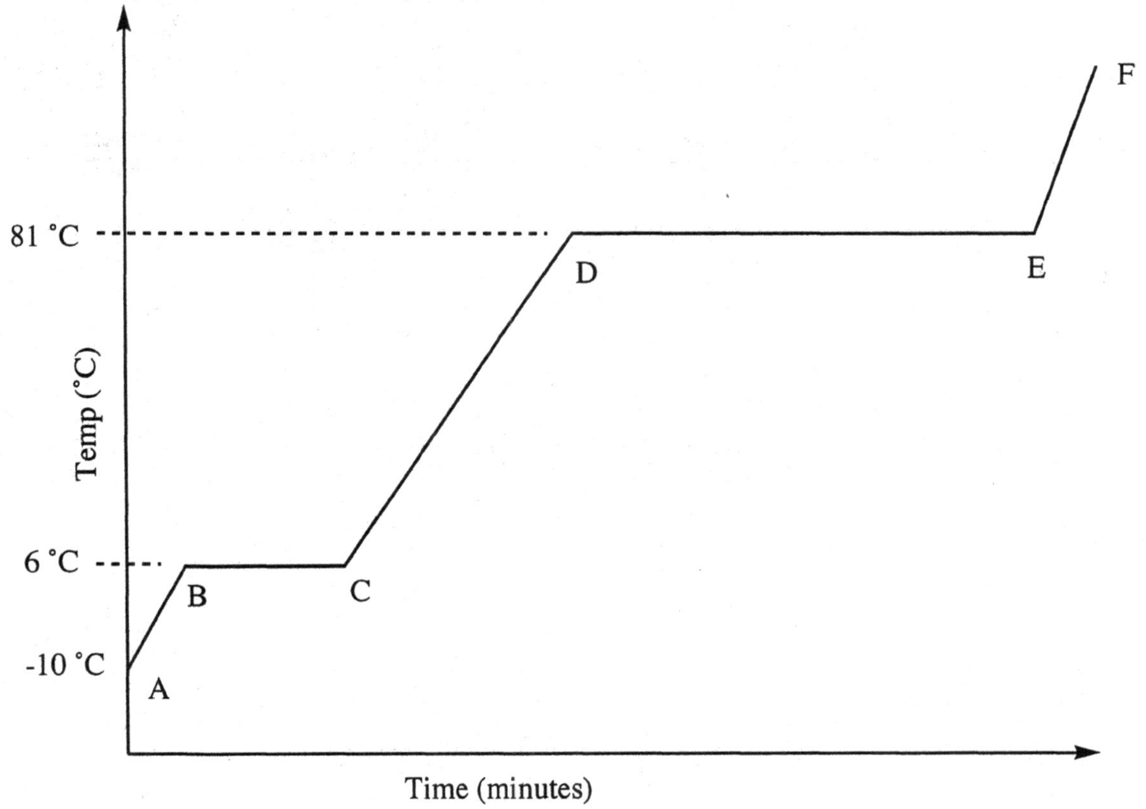

Critical Thinking Questions

1. What is happening to the cyclohexane between points A and B on the diagram?

2. a) What is happening to the cyclohexane between points B and C on the diagram?

   b) Between points B and C, heat is still being added, but the temperature is not changing. What is the heat being used for?

3. What is happening to the cyclohexane between points C and D on the diagram?

4. a) What is happening to the cyclohexane between points D and E on the diagram?

   b) Between points D and E, heat is still being added, but the temperature is not changing. What is the heat being used for?

5. What is happening to the cyclohexane between points E and F on the diagram?

5a. Use your answers to questions 1-5 to annotate each region of the heating curve.

6. We can say that 6 °C is the ____________ of cyclohexane, and 81 °C is the ___________ of cyclohexane.

7. Based on this diagram, what two basic processes can happen as a result of adding heat to a substance?

8. Can you think of other things that may occur if heat is added to a substance?

9. In the closed containers below, draw microscopic representations of the cyclohexane sample at each temperature. Each cyclohexane molecule can be represented by a hexagon: 

At -10 °C   At 40 °C   At 100 °C

Phase ____________   ____________   ____________

10. Above, list the name of the phase of the cyclohexane at each temperature.

11. Above, on the arrows, list the names for each phase change.

12. Whether at 8 °C (just above its freezing point) or at 79 °C (just below its boiling point), cyclohexane is still in liquid form. At a microscopic level, what differences would there be between liquid cyclohexane at 8 °C and liquid cyclohexane at 79 °C? (Hint: think 'dynamically')

Model 2: Some Quantitative information about cyclohexane

Heat capacity of liquid cyclohexane = $1.85 \ \frac{J}{g \cdot C}$

Heat of vaporization of cyclohexane = $393 \ \frac{J}{g}$

Definitions:

Heat Capacity is defined as the amount of energy it takes to raise the temperature of 1 gram of a substance by 1 °C.

Heat of Vaporization is defined as the amount of energy it takes to convert 1 gram of a substance at its boiling point from liquid to vapor (gas).

Critical Thinking Questions

13. Does the transformation below require heat to be input, or does it release heat?

$$C_6H_{12\ (l,\ 6\ ^\circ C)} \longrightarrow C_6H_{12\ (l,\ 81\ ^\circ C)}$$

14. Does the transformation below require heat to be input, or does it release heat?

$$C_6H_{12\ (l,\ 81\ ^\circ C)} \longrightarrow C_6H_{12\ (g,\ 81\ ^\circ C)}$$

15. How much energy (how many Joules) would it take to warm a 100 gram sample of liquid cyclohexane from 6 °C to 81 °C? Show your calculation, with units.

16. How much energy would it take to vaporize (turn from liquid to gas) a 100 gram sample of liquid cyclohexane (assume it is already at 81 °C and doesn't need to be warmed up first)? Show your calculation, again with units.

17. Explain, using your answers to CTQ's 9 and 12, why the numerical answers to CTQ's 15 and 16 are so different.

18. The *heat of fusion* is defined as the amount of heat (in J) that it takes to transform 1 gram of a substance from a solid to a liquid at its melting point.

a) Write a reaction (similar to what you see in CTQ's 13 and 14) that depicts this transformation.

b) An experiment showed that it takes 8.0 kJ to transform a 250 gram sample of cyclohexane from solid to liquid. Calculate the heat of fusion of cyclohexane (in J/g).

Exercises

This problem is to be done as HOMEWORK, to be turned it by each student individually. Do it on a separate page and hand it in in class next Wednesday. Show all your work!

This is a long problem which must be done in stages. If you can draw the appropriate heating curve, conceptualize what is happening at each stage, and do all the calculations, then you clearly understand how to use heating curves.

<u>Substance X has the following thermochemical properties:</u>
Melting point: 20 ˚C
Boiling point: 140 ˚C

Heat capacity of solid X = $2.5 \dfrac{J}{g \cdot C}$

Heat capacity of liquid X = $3.5 \dfrac{J}{g \cdot C}$

Heat capacity of gaseous X = $2.0 \dfrac{J}{g \cdot {}^{\circ}C}$

Heat of fusion of X = $50 \dfrac{J}{g}$

Heat of vaporization of X = $400 \dfrac{J}{g}$

1. Draw a heating curve (temp vs. time) similar to that shown in Model 1. On the curve, clearly indicate the melting point and boiling point of substance X. Show where substance X is a solid, liquid, and vapor. At each part of the curve, also describe what the addition of heat is doing (example "warming the liquid" or "melting the solid" – see CTQ 5a).

2. We are going to calculate the TOTAL amount of energy (in J) required to convert a **500 gram** sample of X from a solid at 0 ˚C to a vapor at 200 ˚C. You can't do this calculation all at once, because the phases keep changing.

a) Describe what is happening at each point on your curve:
   i. Between 0 ˚C and 20 ˚C
   ii. At 20 ˚C
   iii. Between 20 ˚C and 140 ˚C
   iv. At 140 ˚C
   v. Between 140 ˚C and 200 ˚C.

b) For **each** of the five regions of the curve (outlined in question 2a), calculate the amount of heat (in J) to accomplish the transformation. Add these together to get the overall total amount of energy required to go all the way from 0 ˚C to 200 ˚C.

CHEMISTRY 100: ChemActivity 12
Activation Energy

Group Name ____________________ Recorder ______________

Because of the number of surveys conducted today, we will be doing only a very short ChemActivity which deals with a basic understanding of why some reactions happen more readily than others.

Model 1: A rubber ball on an inclined ramp

Observe what happens when a ball at the top of an inclined ramp is released.

Critical Thinking Questions

1. Describe what happened.

2. Where does the ball have a higher potential energy, at the top of the ramp or at the bottom of the ramp?

3. Does this explain why the ball will go from the top to the bottom by itself, but not from the bottom to the top?

Model 2: A rubber ball on an inclined ramp with a bump in it.

Critical Thinking Questions

4. Does the presence of a barrier on the ramp change the answer to question 2? In other words, does the difference in potential energy at the top and bottom of the ramp change if there is a barrier introduced? Why or why not?

5. Why doesn't the ball roll by itself from top to bottom? What must we do to help it 'overcome the barrier'?

Model 3: A "Reaction Energy Profile"

Below is a hypothetical energy diagram for a reaction that converts compound "A" to "B". The horizontal lines shown represent the relative energies of compounds A and B. Their positions is analogous to the rubber ball on the ramp.

A → B

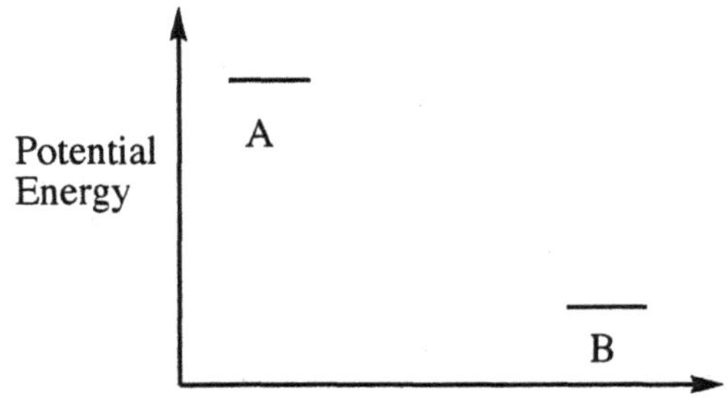

Critical Thinking Questions

6. Which has a higher potential energy, A or B?

7. Which is more stable, A or B?

8. Connect the energies of A and B by a straight line. If you now put a rubber ball at position A, what would happen?

9. Assuming that what you drew in CTQ 8 is the true energy profile, could you ever have a stable sample of compound A at all? Why or why not?

10. Redraw the line connecting compounds A and B: this time introduce a 'barrier' between them.

11. Could the presence of this barrier help explain why a sample of compound A could be stable, even though transforming into B is favorable?

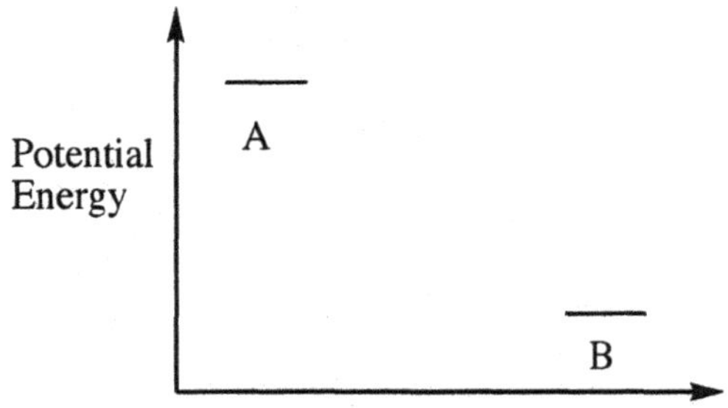

12. How could you overcome this barrier and allow compound A to be transformed into compound B?

Definition: the height of the barrier represents the "activation energy" of a reaction. Reactions with very low activation barriers are very fast. Reactions with high activation barriers are very slow.

Exercises

1. a) The reaction below has a large change in potential energy (downhill) but a very small activation barrier. Sketch a reaction energy diagram that shows this.

$H_3O^+{}_{(aq)} + OH^-{}_{(aq)} \rightarrow 2\ H_2O$

b) Would this reaction be relatively fast or relatively slow? Explain?

2. a) This reaction has a massively larger change in potential energy (also downhill) but also a massive activation barrier. This is the combustion of methane, the reaction that is happening when a Bunsen burner is burning. Sketch its reaction energy diagram.

$CH_4 + O_2 \rightarrow CO_2 + 2\ H_2O$

b) How does your diagram explain:
i. That this reaction produces a lot of heat.

ii. That a Bunsen burner won't start burning until you light a match and bring it nearby.

CHEMISTRY 100: ChemActivity 13:
Electrochemistry

GROUP NAME ___________ RECORDER ___________

One specific type of reaction involves the transfer of <u>electrons</u> from one atom, molecule, or ion to another. We call these electron-transfer reactions or oxidation-reduction reactions, and the field of chemistry that deals with these types of reactions is called electrochemistry.

Model 1

The instructor should have four samples at the front of the room: some copper wire, some solid copper nitrate, some zinc wire, and some solid zinc nitrate. Watch as he attempts to dissolve each of the four samples into water.

Critical Thinking Questions

1. Complete the table below

	Initial appearance of solid	Is it water soluble?
Cu		
$Cu(NO_3)_2$		
Zn		
$Zn(NO_3)_2$		

2. For the samples which dissolved in water, show what species are actually present in solution. Also indicate the color of the solution.

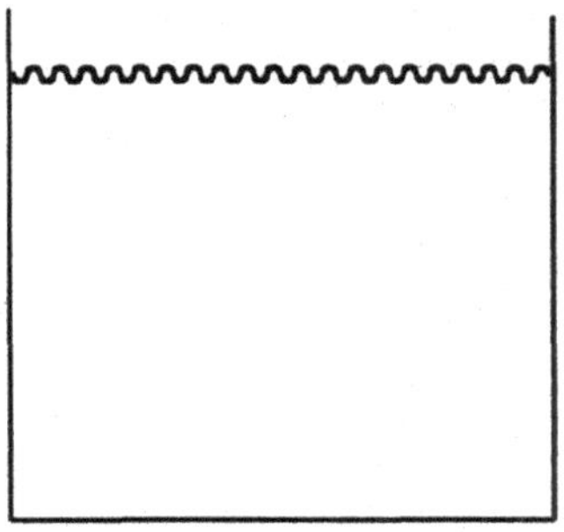

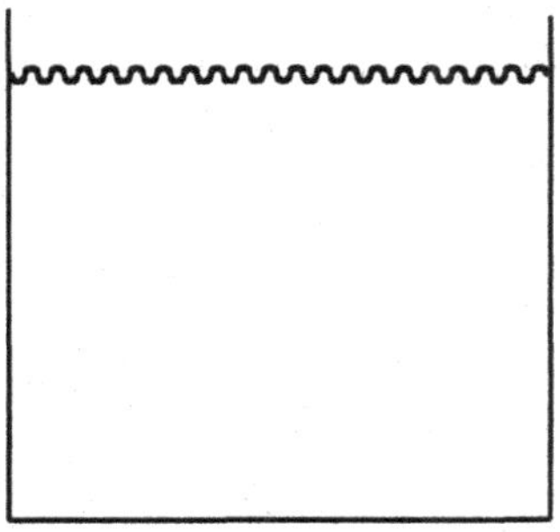

3. Fill in the following table. Use the periodic table to help you.

	Number of protons	Number of electrons	Is it water soluble?
Cu			
Cu^{2+}			
Zn			
Zn^{2+}			

4. What property do copper atoms and copper ions have in common? What properties are different?

5. Given your answer to question 4, do you expect Cu and Cu^{2+} to participate in chemical reactions in the same way? Why or why not?

Model 2: Half-Reactions

$Cu \rightarrow Cu^{2+} +$ ______	$Cu^{2+} +$ ______ $\rightarrow Cu$
$Zn \rightarrow Zn^{2+} +$ ______	$Zn^{2+} +$ ______ $\rightarrow Zn$

Critical Thinking Questions

6. In what sense are the four half-reactions above not balanced?

7. In the blanks, add what needs to be added in order to balance the reactions.

Definitions

When an atom, molecule, or ion loses electrons, we say that it undergoes **oxidation**.
When an atom, molecule, or ion gains electrons, we say that it undergoes **reduction**.

8. Label each half-reaction in Model 2 as either oxidation or reduction.

Model 3: An oxidation-reduction reaction.

Watch as the instructor combines the Cu wire with the Zn^{2+} solution, and the Zn wire with the Cu^{2+} solution.

Critical Thinking Questions

9. What did you observe? Did both combinations results in a reaction?

10. For the one combination which worked, write the half-reactions which correspond to the changes they are undergoing (hint: both half-reactions will be from Model 2).

11. Add up the two reactions to form the overall reaction. If some things appear on both the left and right side of the reaction, you can 'cancel them out'.

Half reaction #1:

Half reaction #2: + ________________________________

Overall combined reaction

12. Which of the starting materials got oxidized? How can you tell?

13. Which of the starting materials got reduced? How can you tell?

14. Complete the sentence. In this reaction, electrons were transferred from the _________ to the _________.

15. Based on the results of the demonstration, which pair of species (listed below) are more <u>stable</u> (lower in potential energy)? Explain your answer.

A. Cu and Zn^{2+} B. Zn and Cu^{2+}

Model 4: A basic battery

While it's nice to be able to do an oxidation-reduction reaction in a beaker, it would be better if the electrons being transferred could be run through a wire. This would produce electric current and could operate an electric appliance. A device called a battery contains all the reagents needed to allow an oxidation-reduction reaction to happen and have the electrons flow through a wire and operate an electric device. A basic schematic of a laboratory battery is shown here.

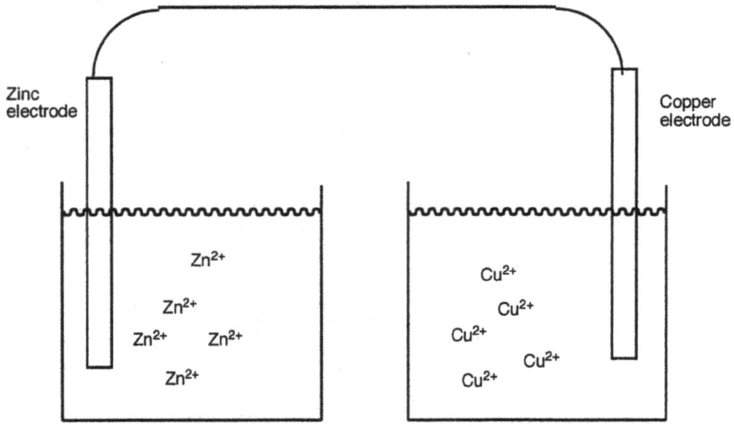

Critical Thinking Questions

16. Again write the correct half-reactions for the oxidation-reduction reaction which occurred when Cu^{2+} reacted with Zn (should be the same answer as question 10). Write them next to where those reactions would take place in the battery.

17. Since the Cu^{2+} and the Zn are not in direct contact, how can this reaction still happen in this system? (Note: only electrons can flow through the wire, not ions or anything else). Explain your answer.

18 Show which direction the electrons flow through the wire to make the reaction work.

19. There is a problem with the setup: you will get a buildup of negative charge at the side where the electrons are going. In order to avoid this problem, a "salt bridge" (which allows very slow passage of ions through a gel) must be added to the circuit. Draw one on the diagram.

20, Over time, one of the metal electrodes will get smaller and the other one will get bigger. Which will grow and which will shrink? Explain your answer.

Exercises

In all cases an oxidation-reduction reaction results in the transfer of electrons from one atom, ion, or molecule to another. The overall reaction will never show 'extra' electrons either on the left or right side of the arrow.

1. Another pair of half reactions that can be used to complete an oxidation-reduction process (or build a battery) are:

$Ag^+ + e^- \rightarrow Ag$
$Ni \rightarrow Ni^{2+} + 2\,e^-$

How can you combine these together to form an oxidation-reduction reaction **with no electrons** appearing on either side of the final equation?

2. Here's another pair of half reactions.

$Cl_2 + 2\,e^- \rightarrow 2\,Cl^-$
$Na \rightarrow Na^+ + e^-$

a) Which is the oxidation half-reaction? Which is the reduction half-reaction?

b) Combine them together to form an overall reaction (**without extra leftover electrons**)

c) Watch http://www.youtube.com/watch?v=Mx5JJWI2aaw to see this reaction in action.

3. Read section 9.1 of the online textbook
http://preparatorychemistry.com/Bishop_Atoms_First.htm

CHEMISTRY 100: ChemActivity 14
Drawing Organic Molecules

RECORDER _______________ GROUP NAME _______________

Today we will look at a very large number of organic molecules, and get practice drawing them, figuring out different ways to arrange the atoms.

Model 1: A pile of organic molecules, drawn in Lewis structure (expanded form)

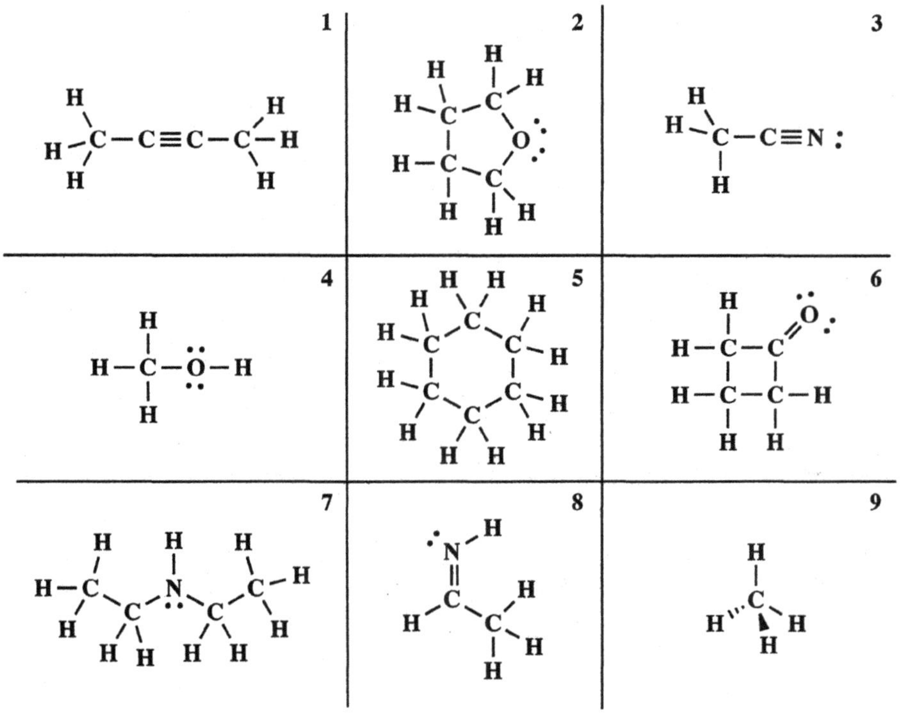

Critical Thinking Questions:

1. Look at all of the carbon atoms that are shown in the molecules above. What do you notice about the number of bonds that each one forms?

2. In the pile, there are THREE ways that each carbon atom can form its required number of bonds. List those ways, and in each case, choose a molecule from the pile which exhibits that type of bonding. An example is shown:

1. Four single bonds
(see molecule 4)

3. There is one way in which carbon can make its required number of bonds which is not represented by any of the molecules in the pile. Can you think of what it could be? Can you draw an example?

4. Use the molecules in the pile to complete these sentences:

   Each oxygen atom forms ______ bonds and has ________ lone pairs.

   Each nitrogen atom forms _______ bonds and has _______ lone pairs.

   Each hydrogen atom forms _______ bonds and has _______ lone pairs.

5. Show two different ways in which oxygen can form its required number of bonds. In each case, show one molecule from the pile which shows that type of bonding.

6. Show three different ways in which nitrogen can form its required number of bonds. Show one molecule from the pile which shows each type of bonding.

7. Some of the molecules in the pile are "hydrocarbons". Mark them with a big **X**.

8. Using the rules that were presented on Monday, draw each of the molecules in the pile in the zigzag shorthand form.

1	2	3
4	5	6
7	8	9 (this one doesn't really work in shorthand... why not?)

Model 2: Structures and Molecular Formulas:

Hint: this last compound has 4 carbons!

Structure:

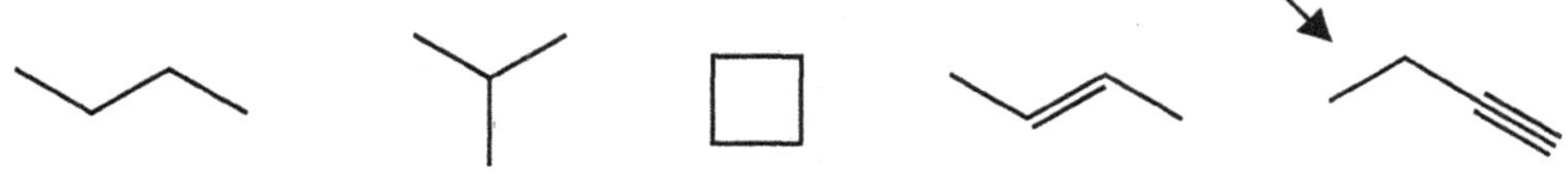

Expanded Form:

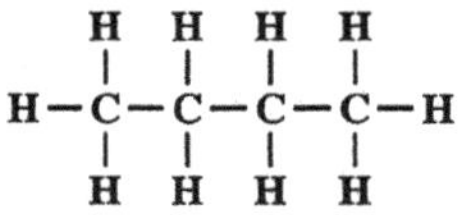

Molecular Formula: C_4H_{10}

Critical Thinking Questions:

9. Count the C and H atoms in the first structure to be sure there are 4 C's and 10 H's.

10. For each of the other structures, draw them in the expanded form and determine their molecular formulas.

11. Indicate any pairs of molecules which are "isomers" (remember the definition of isomers from LabActivity 13) by circling them together.

12. Can you draw structures that follow the rules of bonding (C with 4 bonds; H with 1) that has a formula of C_4H_{11}? C_4H_{12}? If yes, show examples. If no, explain why not. *Hint: you may be able to use the modeling kits to help answer this question.*

13. A hydrocarbon which contains the MAXIMUM possible number of hydrogen atoms is called a **saturated compound.** Which compounds from the above table are saturated? What differences do you notice between the saturated compounds and the other compounds?

14. A hydrocarbon which contains TWO H's less than the maximum number of H atoms has "one degree of unsaturation". Which compounds in the table fit this description?

15. How many other isomers (different from the ones shown above) can you draw that have the formula C_4H_8? Draw them all. Look at the table for ideas on what ways the molecule can have "one degree of unsaturation". *Hint: there are three other answers.*

Note: A good way to answer CTQ 15 is to use the molecular modeling kits (which I brought to class today). Choose 4 black balls (C) and 8 white balls (H) and connect them in such a way that:

- There aren't any 'holes' in the atoms that aren't plugged up by one end of a bond
- You can't take the model and rotate it to make it look like another structure you've already drawn.

16. C_4H_6 has FOUR H's less than the maximum: we say it has "two degrees of unsaturation". How are the number of degrees of unsaturation related to the number of H atoms in the hydrocarbon?

17. Look at the structures in the box in Model 2 and fill in the blanks: Each _______ or _________ counts as one degree of unsaturation. A _______ counts as two degrees of unsaturation.

18. Draw three other isomers (besides the one in the Table in Model 2) that have the formula C_4H_6. *Hint: there is one in the big pile on page 1.*

Exercises:

1. Draw two isomers which each have the molecular formula C_2H_6O. Make sure your drawings follow the HONC rule, and that they are different (that you can't convert one to the other by rotating around single bonds).

2. In the pile of molecules on page 1, there is a compound which has the formula C_4H_8O.
   a) Which one is it?

   b) How many degrees of unsaturation does it have? ______ *Hint: see CTQ 17.*

   c) Draw two more different isomers with this formula.

3. Are hexane and cyclohexane (shown) isomers? Why or why not?

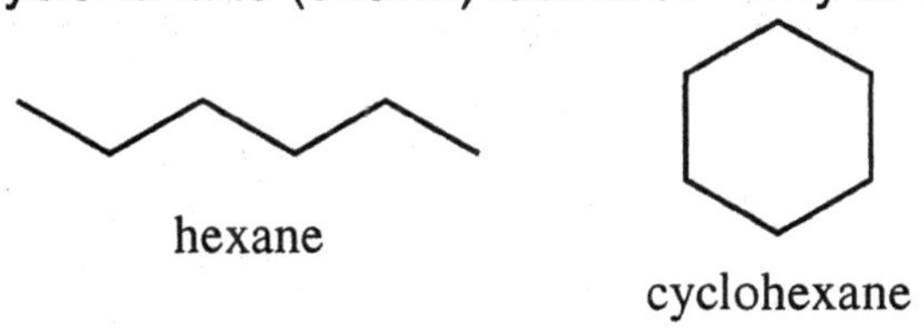

CHEMISTRY 100: ChemActivity 15:
Chirality

GROUP NAME _______________ RECORDER __________

Today we will be looking at one important aspect of organic molecules called *chirality* and how it impacts the interactions of certain molecules with one another.

You should have received a small kit which contains 10 atoms and 8 bonds, along with a few other items.

Model 1: Your hands

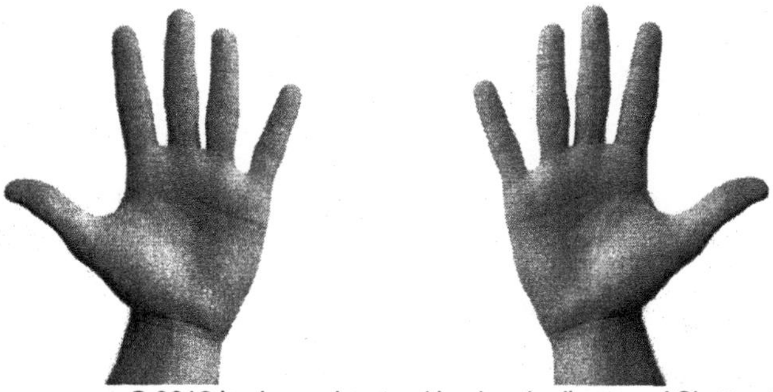

Critical Thinking Questions:

1. In what ways are your hands the same?

2. In what ways are they different (ignore minor differences, jewelry, etc)?

3. How can you describe the geometrical relationship between your right hand and your left hand?

Model 2: A fork

Critical Thinking Questions:

4. Try to hold the plastic fork in your left hand, and then in your right hand. Can you hold it in the same way with either hand?

Model 3: A glove

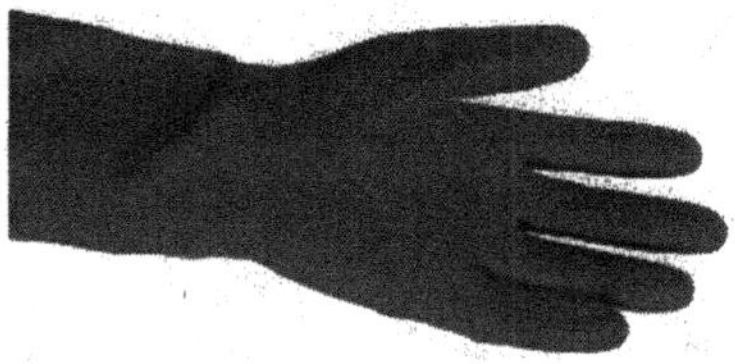

Critical Thinking Questions:

5. Try to put the glove on your left hand, and then your right hand. Can you put it on successfully in the same way with either hand?

6. What property does the fork have (and the glove lack) that accounts for the different answers to CTQ 4 and CTQ 5?

Definitions:

- Any object whose mirror image cannot be fully superimposed onto the original object is a **chiral** object.
- Any object whose mirror image CAN be fully superimposed onto the original object is an **achiral** object.
-

Critical Thinking Questions:

7. Use two forks (borrow one from another group) to prove to yourselves that the mirror image of a fork is identical with the original fork – they can be superimposed.

There is another test for chirality besides the superimposition of mirror images: if an object has INTERNAL MIRROR SYMMETRY, then it is an **achiral** object. Those always go together:

- If an object has internal mirror symmetry, then its mirror image will be superimposable onto the original, and the object is <u>achiral</u>.
- If an object has no internal mirror symmetry, then its mirror image will NOT be superimposable onto the original, and the object is <u>chiral</u>.

Critical Thinking Questions:

8. Show by drawing a dashed line that the fork below has internal mirror symmetry.

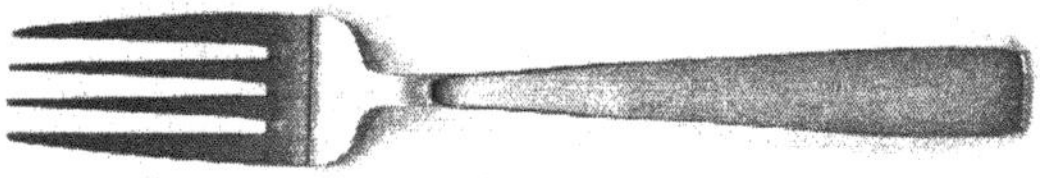

9. Below are some images of household items. For each item, determine whether it is chiral or achiral.

Blank coffee cup

Coffee cup with happy face

Roller skate

Chair

Football (no logo)

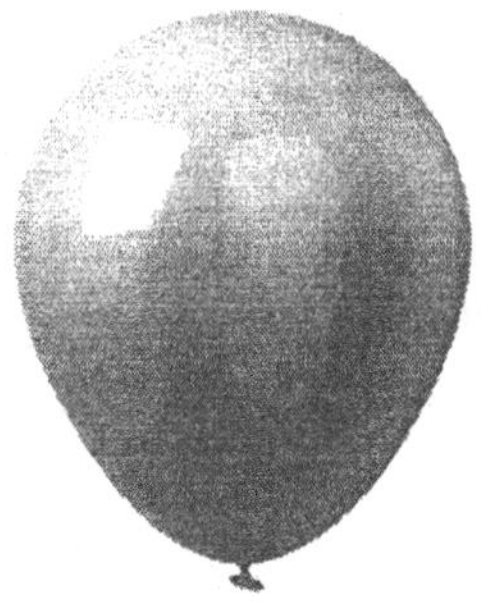

Balloon (blank)

Balloon with "W" on it

Balloon with "Z" on it

Balloon with "3" on it

Model 4: Carbon atoms with 4 different groups attached.

Using the mini-model kits provided, construct one molecule with a carbon atom (black) in the center, and to it connect bonds to 4 different colored atoms: white, green, red, and the other one (some groups have blue as their 4th color, others have silver or purple)

Now, construct a second model with the same arrangement.

Line up the two models facing each other so that they look like mirror images. If you can't do that, you may have to take one of your models, pull off two of the non-black atoms and switch their places. Once you think you have the models constructed properly, **alert the instructor or T.A.** that you are ready so he/she can see if you've actually built mirror images. *Keep these for use with Model 5 as well.*

Critical Thinking Questions:

10. Can you superimpose the two models? __________

11. Based on your answer to question 10, is a tetrahedral C atom with 4 different things attached chiral or achiral? Explain your answer.

Model 5: A Hormone-Receptor relationship

Why should we worry about whether or not various organic molecules are chiral? It is because life is constructed out of chiral molecules: the amino acids and carbohydrates which are used to make our enzymes and our food (from which we get energy) to carry out day-to-day biological processes are almost all chiral molecules. By an accident of biological history, all of the amino acids and carbohydrates in nature are of one mirror image form and not the other.

One group member volunteers his/her **left** hand: draw the letters R (for red), W (for white) and G (for green) on his/her fingers like this:

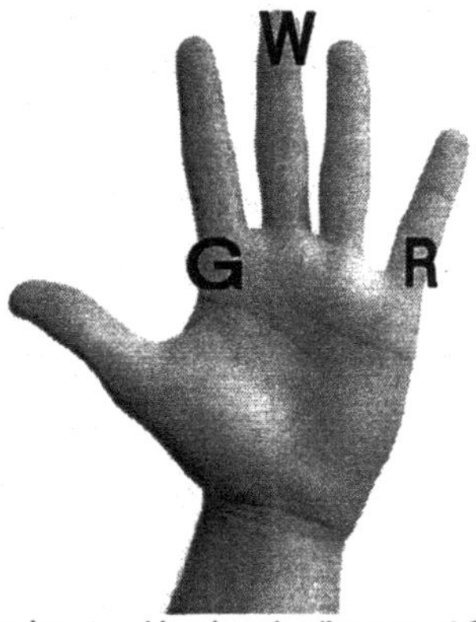

We are going to use the models we constructed in Model 4 to represent a hormone, and your labeled hand we'll pretend is a hormone receptor. Let's say (for instance) that if the hormone can properly fit into the receptor, a biological process is initiated that makes you happy. However, if the hormone cannot properly fit into the receptor, the process is not initiated, and you are not made to feel happy.

Critical Thinking Questions:

12. Attempt to sit one of the two molecular models (from Model 4) on the labeled hand such that the white, green, and red balls are sitting on your finger on the appropriate matching letters. Did it work? If the model is a hormone and your hand the receptor, did this interaction induce happiness?

13. Attempt to sit the other molecular model on the hand with the white, green, and red balls sitting on the correct letters. Did it work that time? Was the biological process leading to happiness initiated or not?

14. In your own words explain the different results you got for the above 2 questions.

15. What do you think you would observe if a test subject consumed a synthetic hormone which...

    a) consisted of only molecules of the correct mirror image (the one which properly fit into the receptor?

    b) consisted of only molecules of the incorrect mirror image – the one which does not properly fit into the receptor?

    c) consisted of a 50:50 mixture of molecules: half of each mirror image?

Exercises

A famous case of one mirror image of a compound interacting with human biology in a positive way and the other mirror image interacting in a negative way is the drug Thalidomide

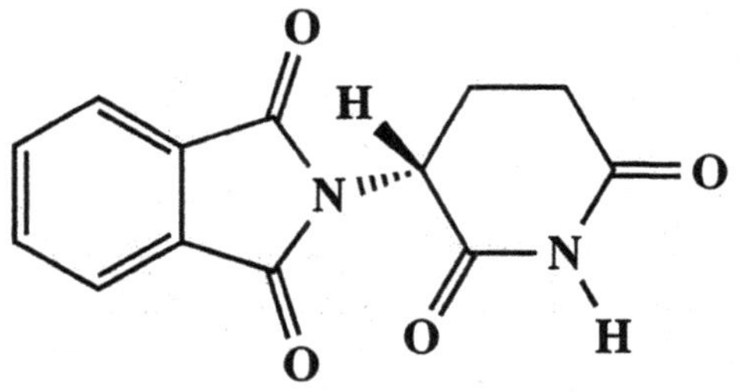

Thalidomide

1. Indicate which carbon atom is responsible for its chirality. What you need to find is one carbon atom which has 4 different atoms (or groups of atoms) attached to it).

2. Do a search on the internet for "thalidomide" and "chirality" or "racemic" to learn the different physiological effects of the two mirror images of thalidomide.

CHEMISTRY 100: LabActivity 1**
Introduction to POGIL and Estimation

Welcome to your first CHEM 100 lab! Today we will learn about the POGIL Activities, meet and learn about your new group members, and do a few exercises in estimation.

General Information about POGIL

POGIL stands for **P**rocess-**O**riented **G**uided **I**nquiry **L**earning. Guided Inquiry, sometimes called "Discovery-Based learning", is a non-lecture approach to classroom instruction which has been shown to increase student learning in the subject and improve students' attitudes towards class. Incorporating a Process-Oriented approach, where students each have a defined role on a team, helps build skills like problem solving, communication, teamwork, management, and assessment: skills future employers are looking for.

Hopefully you have all now received and are wearing your name tags with your individual code number on them. WRITE DOWN THIS NUMBER SOMEWHERE and REMEMBER IT! Your number determines which group you are in and what role you are playing each day. For example, those of you with the same number but different letters (e.g. 5A, 5B, 5C, and 5D) form group 5. This group will probably stay together most of the semester, although there may be some minor tinkering with the groups once or twice.

Each day's lab or class

As soon as you arrive at the lab or the Wednesday lecture each day, find the other members of your group and sit together. Then look over any announcements on the board: particularly the list of which group member has which role today. A folder with the day's assignment will be handed to your group at the beginning of class: the group's manager for that day should take out the "manager's instruction sheet" and start by following its instructions. You usually will not need to wait for the instructor's OK to get started on the day's work. If the instructor needs to get your attention during or before the activity, he'll use the silent call to attention (which will be demonstrated on day 1)

In each of the CHEM 100 labs, as well as in the workshops during the Wednesday lectures, you will be working in a group of 3 or 4, and each group member will have a specific role or roles to play that day. The roles will be rotated for each class, so you will get the opportunity to occupy a number of different roles during the semester.

** Adapted from ChemActivity 1, Garoutte, M. P. *General, Organic, and Biological Chemistry: A Guided Inquiry*, 2007.

Roles

These are some of the typical roles that you will be required to assume during the semester, along with the duties for which that that person is responsible:

Manager

- Actively participates in all team tasks
- Keeps the team's focus on the task at hand
- Asks questions to instructor on behalf of the group
- Distributes work and responsibility to group members
- Resolves disputes within the group
- Verifies that all members are participating and understanding the lesson
- Ensures that all the team's reports are submitted to the instructor at the end of the class

Spokesperson

- Actively participates in all team tasks
- Presents results and answers to the class, either verbally or on the board, as requested.

Recorder

- Actively participates in all team tasks
- Keeps the written record of the day's assignment that is handed in to the instructor (other group members may keep copies for their own records, but the recorder's gets handed in and graded each day)

Strategy Analyst

- Actively participates in all team tasks
- Keeps a record of the team's problem-solving strategies and methods
- Identifies areas in which the team is doing well and what needs improvement
- Prepares the Strategy Analyst's report, which is handed in each day.

Technician

- Actively participates in all team tasks
- Operates the calculator or computer used in that day's lesson

LabActivity 1 – Part A Group Introductions and POGIL Roles

Group Name (see Exercise 1) ____________________

Recorder Name ____________________

Model 1: POGIL Roles

Please read the POGIL information and the list of roles and description of duties on the previous two pages.

Critical Thinking Questions (note: all members should discuss these answers together, and the recorder should write down the consensus answers)

1. List one or two primary duties for each of these roles within a group:
   a) The Manager

   b) The Recorder

   c) The Spokesperson

   d) The Strategy Analyst

2. Decide who would be responsible for dealing with these issues if they were to come up during the day's exercise:
   a) Two people start talking about Barry Bonds, instead of chemical bonds.

   b) Two of the group members disagree with the answer that the other two want to put in the report

   c) The instructor says it's time to take a break for the group to reflect the involvement of each member and team cohesion

   d) The instructor wants to know the answer to question 2 (d)

3. What is the purpose of dividing the members of each group into specific roles each day?

4. What is the purpose of rotating the roles from week to week?

5. Why is there a 'strategy analyst' included in each group?

Exercises

1. Come up with a group name to be used throughout the semester. It **must** somehow incorporate your group's **number**. For example, Group 5 might call themselves "Five Alive" (a fruit punch drink popular in Canada)

2. Complete the following table with information about the group's members

Code #	Name	Major	Hometown	Which role would this person be best at and why?	Which role would this person like the least and why?	Special skill (not school-related)

3. As a group, come up with at least 4 answers to the question "What do you hope to accomplish in college?".

1.

2.

3.

4.

4. Imagine that your group is the Hiring Committee of a new chemical company. Your company currently has 20 employees, but you plan to expand it to 40 by the end of the year. List SIX characteristics that your committee will use to screen the applicants for your company.

1.

2.

3.

4.

5.

6.

LabActivity 1 – Part B Estimation

Model 2:

You should have a kit which contains:

- A golf ball
- A meter stick with cm markings

Critical Thinking Questions:

1. Estimate the number of golf balls that could be packed into our lab room. You can assume that the room is empty of furniture, including the lab benches, shelves, and fume hoods. As a group, decide on a 'plan of attack', then carry out the plan.

Describe your plan of attack here (in point form is fine), and put your final answer into the box at the bottom of the page. When complete, the spokesperson will be asked to write the answer on the board and explain how the group got their answer.

Number of golf balls: [ ]

2. Once all the other groups have finished and the different answers put on the board, discuss amongst yourselves whose answer and strategy you liked the best. Write down here which group's answer you think is the closest to the true answer (you can still pick your own group) and explain why.

CHEMISTRY 100: LabActivity 2
Observing Chemical Reactions

In this experiment, you will perform a number of small-scale chemical reactions.

Question: What types of evidence are there for the occurrence of a chemical reaction?

Equipment:

Each group should have a kit which contains:

- 4 Aqueous solutions in dropper bottles. They are:
    - 6 M HCl (hydrochloric acid) CAUTION: corrosive
    - 1 M NaOH (sodium hydroxide) CAUTION: corrosive
    - 1 M $AgNO_3$ (silver nitrate) CAUTION: toxic
    - 1 M $NaHCO_3$ (sodium bicarbonate)
- Zinc foil, cut into little squares
- Copper foil, cut into little squares
- A 24-well microtiter plate

Combine the six different chemicals in every possible combination of two reagents. When using liquid reagents, use one 'squirt' from the dropper of each reagent. When using solid reagents, use one little square of the metal.

Write down (on the following page) your observations for each combination. Be as thorough as you can. It is up to your group to find the most sensible way of conducting the experiment and organizing the results of the different combinations for presentation.

Please inform the instructor when you have completed this part of the experiment.

Note: "Aqueous Solution" means that most of what is in the dropper bottles is water. The rest (5% to 15% for these samples) are the compounds listed above, which have been dissolved into the water. If any reactions are observed, they are in fact caused by the dissolved compounds themselves, not the water.

LabActivity 2 Group Name ___________ Recorder Name _________

OBSERVATIONS

Critical Thinking Questions

1(a). List the different general observations that may indicate that a chemical reaction has taken place.

Critical Thinking Questions, continued.

1(b). In the absence of any color change, solid formation, formation of bubbles, or change in temperature, can we say that no reaction has occurred?

2. If I tell you that the white solid that formed when mixing HCl and $AgNO_3$ was silver chloride (AgCl), could you write a balanced reaction? *Hint: there is only one other product.*

3. If I tell you that the bubbles formed when mixing zinc (Zn) and HCl were hydrogen (H_2), could you write a balanced reaction? *Hint: there is only one other product.*

4. If I tell you that the bubbles formed when mixing HCl and $NaHCO_3$ were carbon dioxide (CO_2), could you write a balanced reaction? *Hint: the other two products have very common chemical formulas that you probably know.*

5. If I tell you that the solid that formed when you mixed Zn and $AgNO_3$ was silver metal (Ag), could you write a balanced reaction? *Hint: there is only one other product.*

(Incidentally questions 2-5 are not just "yes or no" questions: please write out the reaction I'm asking for)

CHEMISTRY 100: LabActivity 3
Density

Group Name ____________________ Recorder Name ________________

In this lab activity we will consider the relationship between volume, mass, and the composition of various materials. There will be three separation "Kits" in the experiment: your group may be doing any one of the Kits first (see the board).

Warm-Up Question (for everyone)

What is the formula for the volume of a cylinder?

Kit A

Equipment:

You should have a container with 6 small cylinders of different color. There should also be a small centimeter ruler in the container.

Critical Thinking Questions:

1. Here is a list of properties. For the set of cylinders in your kit, determine which of the properties are the SAME for all 6 of the cylinders, and which are DIFFERENT. Put them in the correct box below.

color **composition** **length** **mass** **surface area** **volume**

SAME	DIFFERENT

2. Explain your reasoning for why you put "volume" where you did.

3. Explain your reasoning for why you put "composition" where you did

4. Rank the 6 cylinders in order from the lightest to the heaviest (don't use the balance for this exercise – try to judge them by holding them in your hands).

Use their colors to identify the cylinders:
beige **bronze** **dark gray** **golden** **medium gray** **silver**

LIGHTEST					HEAVIEST

5. Now, using an electric balance, accurately determine the mass of each of the cylinders. *USE THE SAME BALANCE FOR ALL SIX* Write those masses with units in the boxes below under the corresponding boxes from the answers to question 4 (even if your above answer had the masses out of order)

mass:					

6. Using the centimeter ruler, determine the volume (in cubic centimeters, or cm^3) of one of the 6 cylinders. Show below what measurements you took and show your calculations:

7. Are all the volumes the same for all of the cylinders? Explain.

8. For each of the six cylinders, calculate the ratio of $\frac{\text{mass}}{\text{volume}}$. Write that ratio (complete with units) into the corresponding boxes below:

mass/volume ratio:

Kit B

Equipment:

You should have a container with 5 cylinders. There should also be a small centimeter ruler in your container, and an accurate balance will be available.

1. Here is a list of properties. For the set of cylinders in your kit, determine which of the properties are the SAME for all 5 of the cylinders, and which are DIFFERENT. Put them in the correct box below.

color composition length mass cross-section area volume

SAME	DIFFERENT

2. Explain your reasoning for why you put "volume" where you did.

3. Explain your reasoning for why you put "composition" where you did.

4. Explain why you put "cross-section" where you did.

5. Rank the 5 cylinders in order from the shortest to the longest.

Use their colors to identify the cylinders:

gray **golden** **silver** **off-white** **white**

SHORTEST LONGEST

6. Measure the length of each cylinder and the diameter of the circle on one of the ends of each cylinder. Write those values (<u>with units</u>) into the corresponding boxes below:

Length Diam.				

7. For each cylinder, calculate the VOLUME (in cm^3), and put it in the corresponding boxes. Show your calculations for ONE of the cylinders in the space here. ↘

Volume				

8. Using an accurate electronic balance, determine the mass of each of the cylinders.

Mass				

Did you notice anything interesting about the masses?

9. For each of the five cylinders, calculate the ratio of $\frac{\text{mass}}{\text{volume}}$. Write that ratio (complete with units) into the corresponding boxes below:

mass/volume ratio:

<u>Kit C</u>

Equipment:

You should have a container with 5 small cylinders of the same color. There should also be a small centimeter ruler in the container.

Critical Thinking Questions:

1. Here is a list of properties. For the set of cylinders in your kit, determine which of the properties are the SAME for all of the cylinders, and which are DIFFERENT. Put them in the correct box below.

color **composition** **length** **mass** **volume**

SAME	DIFFERENT

2. Explain your reasoning for why you put "volume" where you did.

3. Explain your reasoning for why you put "composition" where you did.

4. Measure the length of each cylinder and the diameter of the circle on one of the ends of each cylinder. Write those values (with units) into the corresponding boxes below:

Length

Diam.

5. For each cylinder, calculate the VOLUME (in cm^3), and put it in the corresponding boxes.

Volume

6. Using an accurate electronic balance, determine the mass of each of the cylinders.

Mass

7. For each of the five cylinders, calculate the ratio of $\frac{\text{mass}}{\text{volume}}$. Write that ratio (complete with units) into the corresponding boxes below:

mass/volume ratio:

SYNTHESIS OF IDEAS (to be completed when your group has finished with all three of the Kits)

Concept: The term for the ratio of $\frac{\text{mass}}{\text{volume}}$ is called **density**

1. Based on your results from Kit C, what can you say about the relationship between density and composition?

2. Based on your results from Kit A, what can you say about the relationship between density and mass (at constant volume)?

3. Based on your results from Kit B, what can you say about the relationship between density and volume (at constant mass)?

4. Write a mathematical formula relating density (d), mass (m), and volume (v).

5. Imagine I have a 28.0 g sample of gold with a volume of 1.45 cm^3.

   a) Calculate its density.

   b) If I had a sample with TWICE as much mass..

      i. What would its volume be?

      ii. What would its density be?

CHEMISTRY 100: LabActivity 4:
Solutions and Dilutions

Group Name ____________________ Recorder Name ____________________

In this experiment we will be working with **solutions**. Chemists typically use the word "solution" to refer to a mixture consisting of a small amount of one compound (called the **solute**) which has been dissolved into a larger amount of a liquid (called the **solvent**). In practice the solvent could be any of a number of different liquids (ethyl alcohol, acetone, etc) but in today's experiment our solvent will always be water. We call solutions that have water as the solvent **aqueous solutions**.

In dealing with solutions, a new concept must be introduced called **concentration**. The concentration of a solution tells you quantitatively how much of the solute there is present in a certain amount of solution. We will be using the most common unit of concentration called **molarity**:

Molarity = moles of solute per liter of solution $\left(\frac{\text{moles of solute}}{\text{liters of solution}}\right) = \frac{mol}{L}$ = mol/L = *M*

Note that molarity is a ratio, and can be used as a conversion factor, like you'll see in ChemActivity 4.

Critical Thinking Questions (show your calculations please – with units)

1. If I take 25 moles of alcohol, and add enough water to make up 20 liters of total solution, then what is the concentration (in mol/L) of alcohol in my solution?

2. If I start with 0.050 moles of NaCl, and add enough water to end up with 80.0 mL of total solution, what is the concentration of NaCl in my solution?

3. Imagine that you have 3.00 liters of an aqueous solution of sucrose which has a concentration of 0.60 mol/L.

a) How many moles of sucrose are in your solution?

b) How many moles of sucrose are in 10 mL of your solution? (you may want to convert the 10 mL to liters first).

c) How many molecules of sucrose are in 10 mL of your solution?

4. Scotch is about 40% alcohol (and 60% water). My grandmother liked to drink scotch and water: she liked the taste of the alcohol. However, she couldn't stand the taste of straight scotch: she thought it smelled bad and it irritated her throat.

a) Which has more alcohol in it, a shot of scotch or a shot of scotch to which I have added water and ice? Explain your answer.

b) Which has a higher concentration of alcohol, a shot of scotch or a scotch-and-water? Explain your answer.

c) How does your answer to question (b) explain why Grandma liked scotch and water but hated straight scotch?

d) The term for when an aqueous solution is made less concentrated by the addition of water is called ____________.

Definitions: A **concentrated** solution has a relatively high number of moles of solute per liter of solution. A **dilute** solution has a relatively low number of moles of solute per liter of solution.

Laboratory Exercise

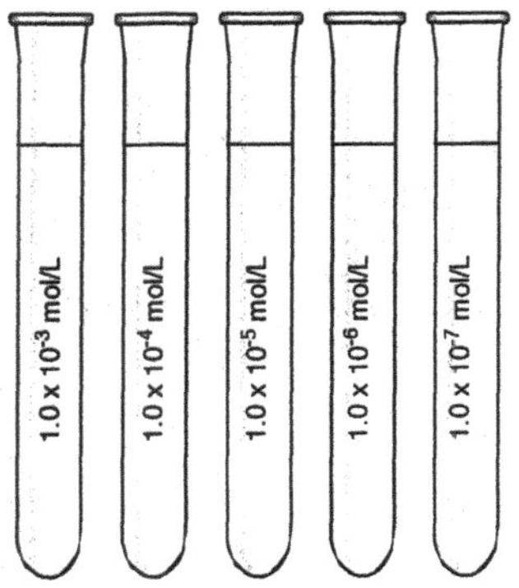

Your group will get about 25 mL each of two solutions:

- a 0.001 mol/L aqueous solution of Yellow #5 (tartrazine)
- a 0.001 mol/L aqueous solution of Red #3 (erythrosine)

Your task is to use the equipment in your drawer to produce a set of 5 solutions for each color (10 test tubes in all). Each test tube should contain approximately 10 mL of solution but each solution's concentration will be different: they must have the concentrations indicated in the figure to the right.

Warm-up Questions:

1. Convert 0.001 mol/L to scientific notation. →

2. Write down below a detailed plan on how your group intends to go about making these solutions. Remember that you only get to use the equipment available in your drawer, although you can clean the glassware and re-use a piece any time you want. Give your plan to your instructor. If he/she is satisfied with it, proceed: if not, modify it to address any issues that he/she noted.

3. When complete, show your instructor your 10 solutions.

Follow-up Questions. In any question involving a calculation please show your work.

1. Describe the color you see in each of the 10 test tubes.

2. Calculate the number of moles of "Yellow #5" in the test tube containing your MOST concentrated solution (assume you have 10 mL in your test tube)

3. Calculate the number of "Yellow #5" <u>molecules</u> in the test tube containing your MOST concentrated solution.

4. Repeat the above calculations for ALL of the other test tubes. You don't need to show your work. Always assume there are 10 mL of sample in each tube.

Concentration:	1.0×10^{-3} mol/L	1.0×10^{-4} mol/L	1.0×10^{-5} mol/L	1.0×10^{-6} mol/L	1.0×10^{-7} mol/L
# of MOLES					
# of MOLECULES					

5. Do any of your samples appear "colorless"? If so, are they colorless because there are none of the colored molecules (Yellow #5 and Red #3) present? Explain.

6. If there still is some detectable color in your most dilute sample, perform additional dilutions (to 1.0×10^{-8} mol/L or 1.0×10^{-9} mol/L) until a colorless solution appears. Recalculate the number of molecules of the solute present and reconsider question #5.

7. Homeopathic medicine** is based on the principle that extremely dilute solutions of the same substances that cause adverse symptoms can be used to treat conditions that cause those symptoms. For example, caffeine can cause insomnia, so a homeopath might prescribe a very very dilute solution of caffeine to treat someone suffering from insomnia. Homeopaths call this the "law of similars".

How dilute is a "very very dilute" solution? One of the most common preparations in homeopathy is a "30C" solution. This means that the original concentrated solution was diluted by a factor of 1:100, and then the new (more dilute) solution was again diluted by 1:100, and so on, until a total of 30 sequential dilutions were done. (the "C" stands for 100)

Let's assume that I start with a **1.00 mol/L** solution of caffeine.

a) Work together to devise and execute a plan to calculate the final concentration of a "30C" preparation of caffeine (in mol/L) and then to calculate the number of caffeine molecules in 1 liter of that solution.

b) Comment on the possible efficacy of a homeopathic remedy prepared to "30C".

** *Note: Homeopathic medicine is commonly confused with herbal medicine. They are not at all the same. If you are given an herbal extract to treat your symptoms, the extract will contain a significant concentration of a number of different compounds. Whether or not these compounds have any effect on your illness is another issue, but at least you are ingesting some substances that may or may not help you.*

CHEMISTRY 100: LabActivity 5:
How much Calcium is there in the Water?

Hard water is water that contains a high concentration of minerals, most notably calcium ions (Ca^{2+}) and magnesium ions (Mg^{2+}). The presence of hard water in a municipal supply can pose problems, including the formation of scale inside pipes and water heaters. Water with a total concentration of calcium and magnesium ions less than 24 ppm is classified as "soft", and that with more than 48 ppm is classified as "hard".

In today's experiment, we will be measuring the concentration of Ca^{2+} ions in a water sample by first carrying out a **titration** of samples of known concentration, plotting a graph, and then analyzing our "unknown". Each group should have brought a sample of water – bottled mineral water, tap water from home, water from the creek – and will report on its calcium concentration.

The reaction that will allow us to determine the level of calcium ions (Ca^{2+}) in water is a complexation reaction that occurs between Ca^{2+} ions and a very large charged molecule called EDTA (ethylenediaminetetraacetate, formula $C_{10}H_{12}N_2O_8{}^{4-}$)

$$Ca^{2+} + EDTA^{4-} \rightarrow [Ca(EDTA)]^{2-}$$

EDTA binds very strongly to a number of ions in solution, including Ca^{2+} and Mg^{2+} (the primary minerals found in hard water) but also poisonous ions like lead (Pb^{2+}) and mercury (Hg^{2+}). If someone is found to have high levels of these ions in their blood, solutions of EDTA can be given intravenously to help eliminate the toxins from the system. This process is called **chelation therapy**. The use of EDTA in chelation therapy for anything other than heavy metal poisoning has not been shown to be beneficial.

General Instructions

Each group will titrate three standardized samples which contain 50, 100, and 150 ppm Ca^{2+}. Each titration must be carried out at least twice. If the volumes of EDTA used in the two trials differ by more than 1.0 mL, you must carry out a third trial.

It is up to your group to determine the best way to present the data for each titration and the results for the experiment. There is a page for your "notes", which won't be graded (can have disorganized material), but on the "Report" page the data should be organized and easy to follow.

On the graph paper provided, plot a graph of Calcium concentration (x-axis) versus the average Volume of EDTA used for that titration (y-axis). Draw a straight line of best fit through your data points (do not just 'connect the dots').

How to do a titration

The process by which one determines the concentration of a solution by adding another solution until an 'end point' is reached is called **titration.** In many cases, including today's experiment, the end point is observed as a color change. In our case, none of the ions involved in the reaction are colored, so we must add an **indicator** to allow us to see the endpoint.

The basic steps to follow for each titration of a sample of Ca^{2+} with EDTA are:

1. Using a volumetric pipet, measure out <u>exactly</u> 50.00 mL of a sample that contains calcium (either a standard or a sample being tested) into a flask.
2. Add 1.0 mL of a solution of sodium hydroxide (6 M) to your sample. Without the solution being very alkaline (having a high pH), the titration doesn't work.
3. Add one small scoop of solid Hydroxynaphthol Blue indicator to the solution. If there are any free Ca^{2+} ions around, the indicator will cause the solution to appear red or pink. If all of the Ca^{2+} has been complexed with EDTA, the indicator will appear blue.
4. Add a magnetic stir bar, and start the sample stirring on a stir plate.
5. Using a buret, slowly add the EDTA solution to the flask with your Ca^{2+} sample in it. Keep adding EDTA until your sample becomes blue. You will need to know and record the exact amount of EDTA solution that was added.
6. After each titration, dump the sample into the sink (don't lose the stir bar) and rinse out your flask two or three times with distilled water. You may then reuse the flask, even if it is still wet.

General Instructions, continued...

Before conducting any titrations on your unknown sample, show your organized data (page 5) and your graph (page 4) to your instructor.

Once your graph is approved, perform the same titration on your sample of bottled water, creek water, or tap water. Carry out this titration at least three times. Use the average volume of EDTA from these titrations and your graph to determine the concentration of Ca^{2+} in your unknown sample.

NOTES (not graded)

REPORT

FINAL RESULTS

Source of Ca^{2+} sample	Calcium concentration (ppm)

CHEMISTRY 100: LabActivity 6:
Preparation of an Ionic Compound

GROUP NAME ____________ RECORDER NAME ____________

All ionic compounds are composed of a combination of positively charged ions (called **cations**) and negatively charged ions (called **anions**). These ions will always be present in ratios that give the compound no overall electric charge – they will be electrically neutral.

The name of any ionic compound is easy to write: it's just the name of the cation followed by the name of the anion. For example:

- K^+ is called "potassium" (same as the metal, which is typical)
- Br^- is called "bromide" (almost the same as the non-metal; the ending 'ide' is typical of anions, as is the ending 'ate')

Since both ions have a charge of 1, they get together in a 1:1 ratio to form the compound KBr, with the name "potassium bromide". The compound formula is also typically written with the cation first and anion last.

As you may have seen in last week's demonstrations, some ions cannot coexist in solution: they get together to form solid **precipitates**, which can be filtered out of the water and isolated.

One example of a pair of ions that form a precipitate is barium (Ba^{2+}) and sulfate (SO_4^{2-}, a complex anion). In today's experiment we will be mixing together solutions that contain free calcium ions and free phosphate ions and collecting the product.

Warm-up Questions

1. How would you write the formula of a compound made of zinc ions (Zn^{2+}) and chloride ions (Cl^-)? Remember the overall charge must be zero.

2. What is the name of the ionic compound from Question 1? ______________

3. Complete the table below by writing the formulas and names of the ionic compounds made by combining the cations and anions shown.

Cation	Anion	Formula	Name
Na^+	S^{2-} (sulfide)		
Mg^{2+}	Cl^- (chloride)		
Ba^{2+}	SO_4^{2-} (sulfate)		
Al^{3+}	OH^- (hydroxide)		
Ca^{2+}	PO_4^{3-} (phosphate)		

Laboratory Exercise

Please read through this entire procedure before starting.

1. Begin by weighing out between 0.90 and 1.10 grams of barium chloride dihydrate ($BaCl_2 \cdot 2\ H_2O$) into a 50-mL or 100-mL beaker. Record the exact amount of barium chloride dihydrate in your Report.
2. Add 10 mL of distilled water and stir until the solid dissolves.
3. In a separate 50-mL or 100-mL beaker, weigh out between 0.90 and 1.10 grams of sodium sulfate. Add 10 mL of distilled water to it and stir until the solid dissolves.
4. Pour one of the solutions into the other one and stir for a few seconds. Your should get plenty of solid barium sulfate.
5. Prepare your equipment for *suction filtration* of your precipitate. You'll need a Glass Fibre** filter paper, your Büchner funnel, and your side-arm Erlenmeyer flask. Set it up how you think it might work, and show your instructor. After getting approval, go ahead and do the filtration. You want to get the maximum amount of solid product into the funnel.
6. If there is any leftover solid in your flask, you may rinse it out with cold distilled water to get additional product. Let the suction filtration continue until there is no water left in the funnel.
7. To fully dry the product, transfer it (removing the filter paper) to a watch glass whose mass you have measured and recorded. Place it in a hot oven for 30 minutes.
8. Remove the hot watch glass with your product from the oven and let it cool.
9. Once the watch glass is cool to the touch, record the final mass of your product, and report it.

** Normal Whatman #1 filter papers won't work for this precipitate: the powder is much too fine and goes through the pores in the filter paper: and your filtrate (the liquid) would appear all milky. The glass fibre filters have much smaller pores, so the solid stays on the filter, and the filtrate is clear.

Report

Masses:

Observations:

Diagram of filtration set-up (with all parts labeled)

Follow-Up Questions – you may want to refer to the ion table on page 1 for help.

1. What is the formula for barium chloride dihydrate?

2. What is the formula for sodium sulfate?

3. What is the formula for barium sulfate (your product?)

4. Draw a balanced reaction showing the conversion of barium chloride dihydrate and sodium sulfate to barium sulfate, water, and another ionic by-product. If you have the formulas wrong in FUQ's 1, 2, and 3, balancing may be impossible.

5. The molecular weight of barium chloride dihydrate is 244.26 g/mol.
   a) How did I know what the molecular weight was?

   b) How many moles of barium chloride dihydrate did you use?

6. The molecular weight of barium sulfate is __________.
   a) How are you going to calculate the molecular weight of barium sulfate?

   b) How many moles of barium sulfate did you make?

7. Explain why the number of moles of barium sulfate made (answer to 6b) cannot be greater than the number of moles of barium chloride dihydrate used (answer to 5b). The balanced reaction in question 4 may help.

CHEMISTRY 100: LabActivity 7:
Debate

GROUP NAME ______________ RECORDER ______________

In today's lab period, each group will be assigned a topic to argue for or against in a debate format. You will have the first 50 minutes of the lab period to use a computer to do research to build arguments for the debate, which will last around 10-12 minutes.

Here is the format we will follow for each debate:

- Opening Statement by the PRO side (2-3 minutes) – Spokesperson #1
- Opening Statement by the CON side (2-3 minutes) – Spokesperson #1
- Short time for sides to prepare their rebuttal statements
- Rebuttal by the PRO side (1-2 minutes) – Spokesperson #2
- Rebuttal by the CON side (1-2 minutes) – Spokesperson #2
- Closing Statement by the CON side (30 sec – 1 min) – Spokesperson #3
- Closing Statement by the PRO side (30 sec – 1min) – Spokesperson #3

Shortly after arriving your group will be given an envelope with your debate topic and your assigned position (either PRO or CON that topic).

If anyone in your group brought a wireless-ready laptop computer, you should be able to access the internet from our lab room. If nobody brought a computer (or if the wireless internet access doesn't seem to be working), there is a bank of internet-ready computers in PHSC 320 that we can use.

At about halfway through the lab period, we will begin the debates. The instructor will serve as moderator. While the other teams are debating, your group will be considering their performance, assigning them a 'score', and deciding on a 'winner'. The winners will not be announced until the whole class is over. Please don't rate your own team on their debate. You can use the judges' report form to help you decide on how to prepare your own arguments for your debate.

The report today will consist of:

- The notes that your two spokespersons are using for their part of the debate (up to 2 pages)
- Your group's evaluation of the other group's performance in the debate.

DEBATE NOTES (page 1)

DEBATE NOTES (page 2)

Judges' Report

Your team will be acting as judge when the other groups are debating. Please leave blank the slot for your own team's debate. For each of the FOUR categories on the scorecard, assign a score from 0 (zero) to 5. Half-points are acceptable. You MUST choose a winner for each debate.

Since you won't be judging your own performance, there are only 2 scorecards.

DEBATE #____ TOPIC ____________________

	Team Name	Opening Statement	Rebuttal	Closing Statement	Use of Supporting Info	TOTAL (max 20)	Winner (✔)
PRO							
CON							

Give some reasons to explain your group's selection of the debate's winner.

DEBATE #____ TOPIC ____________________

	Team Name	Opening Statement	Rebuttal	Closing Statement	Use of Supporting Info	TOTAL (max 20)	Winner (✔)
PRO							
CON							

Give some reasons to explain your group's selection of the debate's winner.

CHEMISTRY 100: Lab Activity 8: Practice Midterm (this is the actual midterm exam from April 2008)

NAME ____________________ GROUP NAME _______________

Test information:

- Time limit: 50 minutes
- Do not start until Dr. Nichols says "GO"
- Time warnings will be given with 15, 5, and 1 minute remaining
- When you hear "STOP" you must stop immediately. Any continued writing will be penalized 5 points.
- You will need a calculator. Dr. Nichols will bring some. You will also need a periodic table. One will be provided (back page)
- For all numerical questions please show calculations *with units.*
- Avogadro's Number = 6.02×10^{23}

---

1. (8 points) The lotto jackpot last week was $71 million (I lost). If I had won that money, how long (in years) would it take me to spend all of it if all I bought was one Starbucks decaf soy mocha (cost $4.45) <u>every minute</u> (all day and all night)? Ignore taxes.

2. (10 points, 4 each except (c) worth 2) Balance the following reactions:

a) $TiCl_4$ + $NaOH$ ⟶ TiO_2 + H_2O + $NaCl$

b) C_5H_8 + O_2 ⟶ CO_2 + H_2O

c) ${}^{222}_{86}Rn \rightarrow \_\_\_\_ + {}^{4}_{2}He$

3. (2 points) What type of nuclear reaction is reflected by reaction 2(c) above? (circle the correct answer – no explanation needed)

a. Fission
b. Fusion
c. α-Decay
d. β-decay

4. (10 points, 5 each) For each of the statements below, an illogical argument is being made. Identify the problem with the argument and provide a short counter-argument to refute it.

a) There must have been aliens on those UFO's that crash landed in Roswell in the 1950's. The only reason we haven't seen the evidence is that the government is hiding it.

b) My grandmother and her friends used to play cards every Thursday night – but they can't anymore since three of them got cancer. I'm going to avoid playing cards too: I don't want to get cancer.

5. (8 points, 4 each) Explain the difference between:

a) A **compound** and a **mixture**

b) **Density** and **mass**

6. (15 points, 5 each) There are multiple choice questions: but you need to provide an explanation for your answer (just choosing the correct answer isn't enough for full credit)

a) How many electrons does a Ca^{2+} ion have?

(a) 18

(b) 20

(c) 22

(d) 40

(e) 42

Explanation:

b) Which diagram most accurately represents a solution of hydrogen peroxide (H_2O_2) in water? (note: the water molecules have not been shown for the sake of clarity).

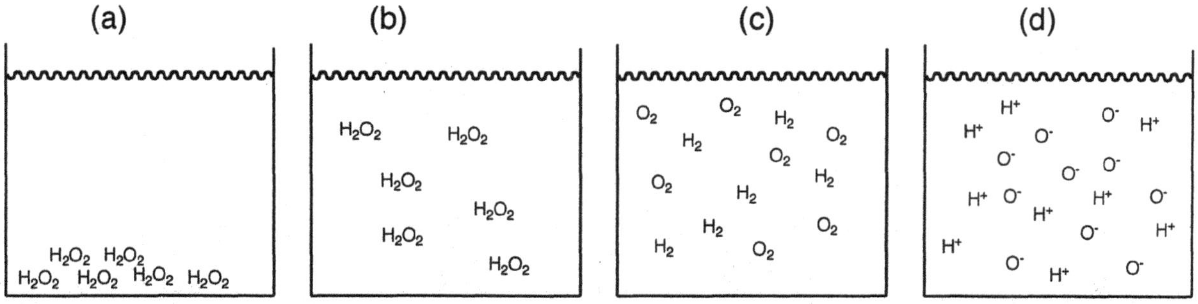

Explanation:

c) If I have a yellow solution that is quite acidic, what should I do?

(a) Drink it! It's probably lemon juice (pH = 2.3)

(b) Drink it! It's probably tea (pH = 5.5)

(c) Don't drink it! It's probably shampoo (pH = 9.0)

(d) Don't drink it! It's probably bleach (pH = 12.5)

Explanation:

7. (19 points, as indicated) Miscellaneous questions:

a, 6 pts) Your RDA for iron (if you are a woman) is 18 mg per day. How many iron atoms are there in 18 mg? *Note: the symbol for iron is **Fe**.*

b, 4 pts) What is the formula and name of the ionic compound made from aluminum ions (Al^{3+}) and thiosulfate ions ($S_2O_3^{2-}$)?

c, 5 pts) Explain what serial dilution means and give an example (you must include numerical values for concentration somewhere in your answer)

d, 4 pts) In terms of subatomic particles, how are the following two things related, and how are they different?

$${}^{79}_{35}Br \qquad\qquad {}^{81}_{35}Br$$

e, 3 pts) What is an "indicator"? (in the context of a titration?)

8. (25 points). Write a 200-300 word mini-essay on **ONE** of the following topics. Your answer will be scored based on the quality of the writing, the use of logical argument, and the correct incorporation of facts and information discussed in class.

a) List some (at least three) possible contaminants found in drinking water and explain which method of purification would be best to use to remove each one.
b) Discuss the scope and the limitations of the use of scientific inquiry.
c) Discuss the advantages and disadvantages of using nuclear power to generate electricity in relation to conventional fossil-fuel based methods of generating electricity.

CHEMISTRY 100: LabActivity 9:
Absorption Spectra of Colored Compounds

GROUP NAME ______________ RECORDER ______________

Today we will revisit the red and yellow solutions with which we did our serial dilutions a few weeks ago. Now we'll look more carefully at why they are the colors they are.

Background

The study of how atoms, ions, and molecules absorb electromagnetic radiation is called SPECTROSCOPY. When electromagnetic radiation (ultraviolet light, visible light, infrared radiation, etc) is shined through a chemical sample, some of the light can be absorbed by the sample, and some of the light will pass through (be transmitted) the sample having not been absorbed. We can use an instrument called a spectrometer to measure exactly how much of the light has been absorbed. A basic set up of a spectrometer is shown below:

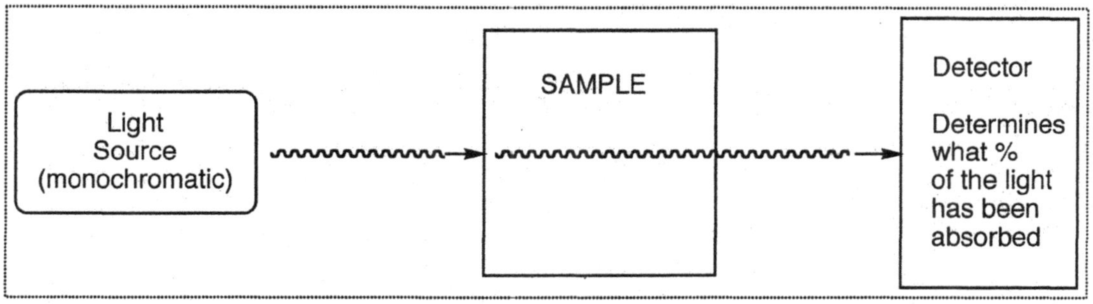

At each wavelength,[1] we can measure the **absorbance** (A) of the light by the sample. Our instrument has a range of absorbance readings from 0 (zero) to around 3. Absorbance is a logarithmic scale:

- An absorbance of 0.00 (A = 0.00) means that 100% of the light was transmitted through the sample (none of it was absorbed)
- A = 1.00 means only 10% of the light was transmitted (90% absorbed)
- A = 2.00 means only 1% of the light was transmitted (99% absorbed)
- A = 3.00 means only 0.1% of the light was transmitted (99.9% absorbed)

The greater the absorbance, the larger the percentage of light at that particular wavelength that is being absorbed by the sample.

[1] The wavelength of light determined what type of radiation it is. Waves of visible light have wavelengths between about 400 and 700 nanometers (nm) – see the rainbow on page 5. Infrared light has longer wavelengths (700 to 10,000 nm). Ultraviolet light has shorter wavelengths (10 to 400 nm)

Part 1: Recording the UV-Vis spectrum of a colored solution

We will be using the Genesys-20 instruments to record the absorbance of our solutions. Before starting, plug it in and turn it on (switch in back) and let it warm up for about 5 minutes. Here is how to use it for each data point:

1. Set the wavelength. To adjust it, use the **▲nm** and **▼nm** buttons.
2. Run a blank to set the zero absorbance. To do this, fill your square cuvette with distilled water, stick it down into the slot, close the lid, and press the **ABS 0/100% T** button. After a few seconds, the display should read **0.000 A**.
3. Record the absorbance of your sample. To do this, remove the blank, insert your sample, close the lid, wait a few seconds until the absorbance reading remains constant, and write it down.

It's important to remember that *you need to run a blank sample each time you change the wavelength!*

You'll need a total of about 25 mL of your colored solution to complete the entire day's lab project:
Odd-numbered groups: please use the yellow solution (1.0×10^{-4} M Yellow #5)
Even-numbered groups: please use the blue-green solution (2.0×10^{-5} M Green #3)

Here's how I would recommend running the full spectrum

1. Prepare two cuvettes: one full of distilled water, the other full of your solution.
2. Start with the wavelength at the lower limit of the instrument (325 nm). Set the blank with distilled water and record the absorbance (see instructions above).
3. Change the wavelength to 340 nm, set the blank again and record the absorbance.
4. Change the wavelength to 360 nm, and do it again. Keep changing it by 20 nm each time until you get to 680 nm, setting the blank and recording the absorbance each time.
5. Plot your results on the graph paper on page 4: wavelength on the long axis (use the whole page: from 320 to 680 nm) and absorbance on the short one. Connect your points with a smooth curve.

Once you have finished your graph, show the instructor for approval.

Once it is approved by the instructor, partner with another group that did the sample of the other color. Use their data to plot the absorbance spectrum of their compound on the SAME GRAPH. Use different colored ink to show the other spectrum; make sure the curves are labeled so the grader knows which curve is for which solution.

DATA for Part 1 (organized)

Follow-Up Questions for Part 1

1. For the yellow solution, determine the λ_{max} (pronounced "lambda-max") from the graph. This is the <u>wavelength</u> at which the absorption curve is the highest.

2. Determine the λ_{max} for the green solution.

3. Shown below is the visible spectrum - the colors of the rainbow- with the associated wavelengths (sorry – no colored photocopies... they're too expensive). Use the rainbow to determine the COLOR of the light that is absorbed to the greatest extent by the yellow solution and the green solution.

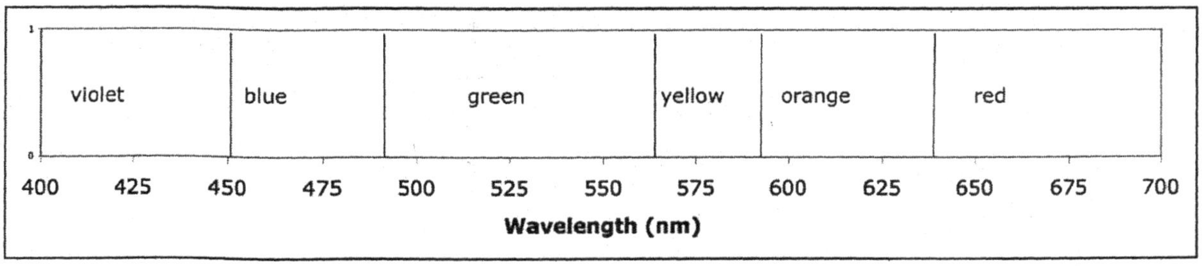

4. Why do you think the yellow solution appears yellow and the green solution appears green?

5. Based on the absorbance of the two samples, predict what color they would be if mixed together in a 1:1 ratio. Explain your prediction.

6. Go ahead and mix equal amounts (say, 1 mL of each) of the yellow and green solutions. What color was it?

Part 2: The relationship between Absorbance and Concentration

For this part of the experiment, you'll need to prepare a series of 6 samples whose absorbances you will measure. You will need about 5 mL of each solution. Use the same colored sample you used for part 1.

The solutions you will need are:

1. A blank (distilled water)
2. A solution whose concentration is 20% of your stock solution.
3. A solution whose concentration is 40% of your stock solution.
4. A solution whose concentration is 60% of your stock solution.
5. A solution whose concentration is 80% of your stock solution.
6. The stock solution (100% concentration).

Before preparing the samples, write up a plan for the preparation of the samples listed above. Show your plan to the instructor for his/her approval before continuing. Part of the plan should include what concentration (in mol/L) each sample will have. The concentrations of the stock solutions are shown on page 2.

Plan:

Here's how I would recommend running the samples:

1. Set the instrument's wavelength to the λ_{max} of your sample.
2. Set the absorbance at zero with your blank sample (see page 2 for instructions)
3. Put each sample into the cuvette and measure and record its absorbance. Between each sample you'll need to clean out the cuvette but you shouldn't need to readjust the zero setting (as long as you don't change wavelengths)
4. Plot a graph of concentration vs. Absorbance on the graph paper on page 8. Draw a calibration curve through your points. Put the absorbance on the short axis of the graph again. Show the graph to the instructor for approval.

DATA for Part 2 (suggestion: use a table with “sample number, concentration, and A)

Follow-Up Questions for Part 2

1. What is the shape of the calibration curve of your graph?

2. What is the relationship between concentration and absorbance?

3. If I gave you a sample of your compound, whose absorbance (at the wavelength you were using) is 1.05 absorbance units, what must its concentration be? Explain how you got the answer.

CHEMISTRY 100: LabActivity 10:
Generation of Hydrogen Gas

GROUP NAME ________________ RECORDER ____________

In today's experiment we will be reacting one known metal (magnesium) and one unknown metal with hydrochloric acid (HCl) in order to generate hydrogen gas.

The reaction we are doing today on the known metal (magnesium) is:

$$Mg_{(s)} + 2\ HCl_{(aq)} \longrightarrow MgCl_{2\ (aq)} + H_{2\ (g)}$$

We will be reacting different masses of magnesium with acid, measuring the volume of hydrogen gas produced, and then determining a relationship between the them.

Safety precautions:

1. **Hydrogen gas is flammable. Please no open flames in the laboratory.**
2. **Dilute hydrochloric acid (1.5 M) is corrosive. Please wash your hands thoroughly if you get some on you. If you spill any, inform the instructor and he/she will clean it up.**
3. **Wear your safety glasses throughout this experiment.**
4. **The 100-mL burets we are using are expensive. Don't break them.**

Technique: The use of a Eudiometer

We need to have a device to measure the quantity of H_2 gas that is produced in each of our reactions. We will be setting up a simple eudiometer using a 400 mL beaker, a J-shaped glass tube, and a closed-end 100-mL buret.

How to set it up (instructor will demonstrate)

1. Fill the 100-mL buret completely with water (to overflowing). Tap water is fine.
2. Put around 250 to 300 mL of tap water into the 400-mL beaker.
3. With your finger over the open end of the 100-mL buret, immerse that end into the 400-mL beaker. Once the open end is underwater, you can remove your finger. If you get any bubbles in the 100-mL buret, you should try this again.
4. The 400-mL beaker rests on the countertop. The buret should be clamped tight enough so it doesn't tip over but loose enough so you can slide it up and down.
5. Put the tail end of the J-shaped tube into the open end of the buret. Put the end with the rubber stopper on the countertop until you need it.

Part 1: Finding a relationship between the mass of magnesium reacted and the volume of H_2 produced.

Read through these directions completely before starting.

1. Measure and record the mass of a piece of magnesium ribbon to the nearest 0.001g. It should have a mass between 0.025g and 0.035g. If it's larger than that, you can cut it with a scissors until it's the right size.
2. Prepare the reaction flask by adding 25 mL of hydrochloric acid (1.5 M) to your 125-mL Erlenmeyer flask.
3. Make sure you have the eudiometer completely ready as described on page 1.
4. Drop the Mg ribbon into the acid. IMMEDIATELY shove the rubber stopper onto your flask and squeeze tight to make sure there are no leaks. The Mg should react vigorously, and the H_2 gas produced should bubble through the J-tube into the 100-mL buret. Wait until no more gas is being produced and the Mg has completely disappeared, and record the volume of gas in the buret.
5. When the reaction is over, dump the water out of the buret and set up the eudiometer again (see page 1) to prepare for the next reaction.

- Once you have carried out the experiment as described above, do it <u>two more times</u> with larger pieces of Mg ribbon:
  - ☆ One should have a mass between 0.055 and 0.065 grams.
  - ☆ The other should have a mass between 0.085 and 0.095 grams.
- If you can't find pieces of Mg big enough, using two strips of ribbon whose combined masses are within the ranges listed above is acceptable.
- You DO NOT have to get fresh hydrochloric acid each time. There should be plenty of acid in that initial 25 mL sample to react with all three of the magnesium samples.

In the space below, write a table which includes:

- The *mass* of magnesium used in each run
- The number of *moles* of magnesium used (calculated), and
- The volume of H_2 gas produced in each run.

DATA:

On the graph paper below, plot a graph with the **number of *moles* of Mg** used on the X axis and the volume of H_2 gas produced on the Y axis.

Follow-up Questions for Part 1:

1. What is the relationship between the **number of moles of Mg** used vs. the volume of hydrogen gas produced? Draw a line on the graph that best fits the data (don't just connect-the-dots) and that goes through the origin.

2. Looking at the reaction that we did, answer true or false for each of the statements below:

   $Mg_{(s)} + 2\ HCl_{(aq)} \longrightarrow MgCl_{2\ (aq)} + H_{2\ (g)}$

_______ a) For every one magnesium atom that reacts, we should get one molecule of hydrogen (H_2) produced.

_______ b) If 1000 magnesium atoms undergo this reaction, we should get 1000 molecules of H_2 gas.

_______ c) If 6.02×10^{23} magnesium atoms undergo this reaction, we should get 6.02×10^{23} molecules of H_2 gas produced.

_______ d) For every 1 mol of magnesium atoms that react, we should get 1 mol of H_2 gas produced.

_______ e) For every 1 gram of magnesium atoms that react, we should get 1 gram of H_2 gas produced.

3. It is impossible to have said "T" to both questions 2(d) and 2(e). Explain, using atomic and molecular masses, why saying "T" for question 2(d) implies "F" for question 2(e).

4. Since the number of moles of Mg used = the number of moles of H_2 produced, what is the relationship between the number of moles of H_2 gas and its volume? Hint: use your graph. Explain your answer.

5. Use your middle data point (the one where the ribbon had a mass of approximately 0.060 grams) to calculate the volume of <u>1 mol</u> of hydrogen gas. Show your calculations and explain how you logically arrived at your answer. *Note: this question may be trickier than it looks. The answer to question 2(d) is very important in understanding why this calculation works.*

Part 2: Determination of the atomic mass of an unknown metal

We will be using another metal that undergoes the same types of reaction with HCl. I'll call it "X". The balanced reaction is:

$$X_{(s)} + 2\ HCl_{(aq)} \longrightarrow XCl_{2\ (aq)} + H_{2\ (g)}$$

Warm-Up Questions:

1. How many moles of H_2 gas would be produced from the reaction of 0.10 mol of metal X? Explain your answer.

2. Here is the reaction for magnesium. How many moles of H_2 gas would be produced from the reaction of 0.10 mol of Mg?

$$Mg_{(s)} + 2\ HCl_{(aq)} \longrightarrow MgCl_{2\ (aq)} + H_{2\ (g)}$$

3. Were your answers to warm-up questions 1 and 2 the same? If so, why? If not, why not?

Because X is a powder, it would be difficult to pour it all into the acid and get the rubber stopper on in time. As such we'll have to use a slightly different technique:

1. Put a small, clean, dry 4-mL vial (provided) onto the balance and set the zero. Then measure out between 0.140 and 0.190 grams of the powder. Record the exact mass you are using. Don't put a lid on the vial.
2. Prepare the reaction flask by adding 25 mL of hydrochloric acid (1.5 M) to your 125-mL Erlenmeyer flask.
3. Make sure you have the eudiometer completely ready as described on page 1.
4. Using tweezers, carefully lower the vial into the 125-mL flask. Don't let it tip over yet.
5. Put the rubber stopper onto your flask and squeeze tight to make sure there are no leaks.
6. Jostle the flask until the vial tips over and acid solution enters the vial. Keep shaking it around to ensure good mixing between the powder inside the vial and the solution around it. Shake the flask periodically until no more gas is released. Record the final volume of $H_{2\ (g)}$ collected.
7. Don't clean things up just yet: you'll have to show the instructor your calculations on the next page: if your results look good you can put your equipment away. If not, you can run this experiment again (you won't need fresh acid).

DATA for Part 2:

Follow-Up Questions for Part 2:

1. Use the graph from Part 1 to answer this question. What number of moles of magnesium would have given you the volume of $H_{2\,(g)}$ which was produced in Part 2? How did you get your answer?

2. Can you say that your answer above is the number of moles of metal X that reacted? Why or why not? (Hint: see the answer to warm-up question 3 on page 5)

3. Here's the important calculation:
a) How many grams of metal X did you use? ____________ (from data)
b) How many moles of metal X reacted? ____________ (from FUQ 2)
c) Show how to calculate the <u>atomic mass</u> of X in grams per mole:

4. Based on your answer to question 3, what do you suppose metal X is?

Show your answers to Questions 3 and 4 to the instructor: if they are close enough to the correct answer, you're done: clean your glassware and return the 100-mL buret to the stockroom.

CHEMISTRY 100: LabActivity 12:
Determination of the Energy in Ethanol

GROUP NAME ________________ RECORDER ____________

Introduction & Background:

In this experiment you will determine the amount of energy (heat) released during a combustion reaction. Combustion, also known as burning, is a chemical reaction between a fuel and some oxidizing agent (like oxygen gas in the atmosphere). The fuel being used in this lab is ethanol, which is common drinking alcohol. When ethanol burns it produces carbon dioxide and water vapor as products, along with thermal energy.

$$C_2H_5OH\ (l) + 3\ O_2\ (g) \rightarrow 2\ CO_2\ (g) + 3\ H_2O\ (g) + \text{ENERGY}$$

We shall measure the amount of heat released by burning ethanol in terms of calories, a common unit of energy. The definition of a calorie is the energy required to increase the temperature of 1 gram of water by 1 degree Celsius. Therefore we can use the heat generated when burning a known mass of alcohol to heat a known mass of water to obtain a quantity of energy. Keep in mind that we are using a somewhat crude calorimeter so some of the heat will escape around the can and not be measured, so you will get an energy value less than the actual value. Nevertheless we can determine an approximate value.

Materials Needed:

Alcohol sample & lamp
Aluminum can (calorimeter)
Hole-punch
Ring stand with 1 large iron ring
400-mL beaker
Digital thermometer
Glass stirring rod

Procedure:

1. Get an aluminum soda can and completely cut the can using scissors about 1 inch from the top. This will serve as our water container. Using a hole-punch, make 2 holes directly across from each another roughly ½ inch from the top of the water container. Clean and dry the inside of the can (*Careful! edges may be sharp!*), then weigh it and **record the mass** in the attached data table.

2. Place a small amount of ice in the beaker and fill it with tap water. Pour the chilled water into the can leaving about ¼ inch space from the holes at the top (there shouldn't be any ice in the can). Weigh the can after adding the water and **record the mass**. Obtain a ring stand and an iron ring in order to set up the calorimeter. Stick the stirring rod through the 2 holes in the can and position it from the iron ring.

3. Get a full alcohol lamp, weigh it and **record the mass**. Place the lamp on the base of the ring stand and lower the soda can so that the wick is nearly touching the bottom of the can (See Figure above). At this time you should take an initial temperature reading (in °C) of the water using the digital thermometer and **record it**.

4. Light the alcohol lamp while continuing to measure the temperature. The water will heat quite rapidly so don't leave the experiment unattended. We don't want the water to get close to boiling because mass will be lost as water evaporates, therefore the temperature of the water should be kept under 60 °C. When the temperature of your water is greater than 50 °C extinguish the alcohol lamp by blowing on it (you might need to take a deep breath). Continue to measure the temperature of the water while using the thermometer to stir, and **record the highest temperature value you observe**.

5. Again, weigh the alcohol lamp and **record the final mass**.

6. Dump the warm water out of the can and dry it out. Then repeat the whole experiment (steps 2-5) again.

REPORT:

Trial 1:

Mass of can with water __________ g
Mass of empty can __________ g
Mass of water __________ g

Initial mass of lamp and ethanol ________ g
Final mass of lab and ethanol ________ g
Mass of ethanol burned ________ g

Maximum water temp __________ ˚C
Initial water temp __________ ˚C
Change in temp (ΔT) __________ ˚C

Heat absorbed by the water: q = mcΔT ($c = 1.00 \frac{\text{cal}}{\text{g} \cdot \text{C}}$) _________ cal

Heat released by the alcohol ___________ cal

Heat released by the ethanol per gram __________ cal/g

Trial 2:

Mass of can with water __________ g
Mass of empty can __________ g
Mass of water __________ g

Initial mass of lamp and ethanol ________ g
Final mass of lab and ethanol ________ g
Mass of ethanol burned ________ g

Maximum water temp __________ ˚C
Initial water temp __________ ˚C
Change in temp (ΔT) __________ ˚C

Heat absorbed by the water: q = mcΔT ($c = 1.00 \frac{\text{cal}}{\text{g} \cdot \text{C}}$) _________ cal

Heat released by the alcohol ___________ cal

Heat released by the ethanol per gram __________ cal/g

FOLLOW-UP QUESTIONS

1. The accepted value for the heat of combustion of ethanol is 7091 cal/g. Compare this to your value. If your value is substantially different from 7091 cal/g, suggest at least 2 possible reasons why. *Note: "human error" and "calculation error" aren't acceptable answers: if you doubt your calculations have the instructor check them.*

2. A standard shot glass is 1 ounce (29.6 mL) in volume, and most hard alcohols are 40% ethanol by volume.

a) Calculate the volume of ethanol in a shot.

b) Find the mass of ethanol in a shot in grams. Let's assume it's vodka and the density is 0.909 g/mL.

c. Calculate the amount of energy in a shot in calories (based on your measured amount of calories released per gram of ethanol burned).

d. Calculate the number of nutritional Calories (Cal) in your shot (1 Cal = 1000 cal).

CHEMISTRY 100: LabActivity 13:
Molecular Geometry

GROUP NAME ________________ RECORDER ______________

Today we will examine the 3-dimensional structure of molecules (atoms held together by covalent bonds) using models. Since atoms themselves are too small to see or manipulate on an individual basis (without huge machines costing millions of dollars) we can use come observed properties of molecules to create plastic models which have some of the correct properties (e.g. molecular geometry) as the molecules would themselves.

Part A: Molecular Geometry

Reading Assignment: Please read Section 3.3 in your textbook for more on molecular geometry.

Each group should have a small kit with purple clay, yellow clay, and several toothpicks. Answer the questions that are included in the instructions on the report form (page 3).

Instructions

1. With your modeling clay, make one purple ball and 6 yellow balls. Try to make the yellow balls approximately the same size, but you can make the purple ball as BIG as you like.
2. The purple ball will be the CENTRAL ball in all of the models you are about to build.
3. Start with 2 yellow balls and the purple ball. Connect them together with the toothpicks with purple ball in the center such that the two yellow balls are as FAR APART as they can possibly get.
   - Question: Draw a picture of what your model looks like
   - Question: Describe the shape of your model: does it look like a letter of the alphabet, a common object, etc?
   - Question: Use your protractor to measure the *angle* between the two bonds.
4. Take your model apart.

5. Now use THREE yellow balls and the purple ball. Again, connect them in such a way that all of the yellow balls are as far away from each other as possible. Remember to keep the purple one in the center: each yellow ball should be connected to the central purple ball.

   - Question: Draw a picture of what your model looks like
   - Question: Describe the shape of your model: does it look like a letter of the alphabet, a common object, etc?
   - Question: Use your protractor to measure the *angle* between any two bonds.

6. After taking your model apart, use FOUR yellow balls and the purple ball. Connect the yellow ones around the central purple ball with the yellow balls as far apart from each other as possible.

   - Question: Draw a picture of what your model looks like
   - Question: Describe the shape of your model: does it look like a letter of the alphabet, a common object, etc?
   - Question: Use your protractor to measure the *angle* between any two bonds.

7. Once you have completed the above sections, let the instructor know (give him the 'thumbs-up' signal). Compare your answers, particularly the last one, with those of the other groups. If some groups have different structures, work together to determine which structure actually has the yellow balls the farthest apart. Fix your original model(s) and answer(s) where necessary.

8. Now use all SIX yellow balls and the purple ball and construct a model with the yellow balls as far apart as possible. Answer the same questions for that model as you did for the others.

9. (HARDER). Do the same as part 8 but with FIVE yellow balls.

REPORT for Part A

Part B: Using Plastic Molecular Models

This week you will check out a molecular model kit form the stockroom and build molecules with them to help with visualization of organic compounds in three dimensions.

Typical colors used in organic molecular models are:

Black	Carbon
White	Hydrogen
Red	Oxygen
Blue	Nitrogen
Green	Chlorine or Bromine
Yellow	Sulfur

When constructing a single bond as part of a molecule, use the short, rigid connectors.

When constructing a double bond or a triple bond as part of a molecule, use the flexible connectors (use **two** for a double bond, and use **three** for a triple bond)

Critical Thinking Questions

1. Build methane (CH_4 – all single bonds), and draw a diagram of your model below. (use the letters C and H instead of balls to represent the atoms in your drawing). Does it look the same as any of the structures you built in part A? Which?

2. What word describes the geometry of methane?

3. Now build formaldehyde – its formula is CH_2O. The C atom is in the center, and from it comes single bonds to two H's and a double bond to an O. Draw a diagram of your model below. Does it have the same basic shape as any of the structures you built in part A? Which?

4. What word describes the geometry of formaldehyde?

5. Now build carbon dioxide (CO_2, both O's are connected to the C by double bonds). Draw a diagram of your model below. Does it have the same basic shape as any of the structures you built in part A? Which?

6. What word describes the geometry of carbon dioxide?

7. Build a model of ethane (C_2H_6).

a) Arrange your model so that can be set on the table with 4 of the H atoms touching the table's surface. Draw what that looks like, using the dash-wedge drawing rules we discussed earlier.

b) Now arrange the model so that it can be set on the table with only **3** of the H atoms touching the table's surface (and not 3 that are all connected to the same C). Draw this structure.

c) How are your diagrams from part (a) and part (b) different? How did you manipulate the model to go from diagram (a) to diagram (b)?

d) Do both diagrams (a) and (b) represent structures of ethane?

8. Build models of these two molecules (Cl can be represented by the green ball).

a) Do the molecules have the same molecular formula?

b) Can you convert from one to the other by doing an internal rotation?

c) Explain the difference in terms of allowed and disallowed atomic motion in compounds built with single bonds and those built with double bonds.

9. Build a model of propane, which has a formula C_3H_8. Draw it in any orientation you like.

10. Now replace one of the H atoms in your molecule with a Cl atom (a green ball) – your new molecule has the formula C_3H_7Cl. Draw a representation of your model.

11. Can you build a model that also has the formula C_3H_7Cl, but has the atoms connected to each other in a different way? (Hint: yes you can!) – build it and draw it below.

12. Can you convert from structure (10) to structure (11) by doing internal rotations? What does that tell you about the two structures?

13. What is the general term for a pair of molecules that have the same molecular formula but have their atoms connected differently?

14. Now build a molecule whose formula is C_4H_{10}. Draw it here.

15. In conjunction with another group (you'll have to consult with them since each kit has only 6 carbons and 12 H's and you need more), construct a different molecule that has the same formula C_4H_{10} but with the atoms connected differently. Draw it.

16. Your structures in answers 14 and 15 represent ________________.

17. Build acetylene (C_2H_2) – make sure that each carbon atom has 4 bonds.

18. What word best describes the basic geometry of acetylene?

CHEMISTRY 100: LabActivity 14:
Candy Chromatography

GROUP NAME____________________ RECORDER________________

This lab has been adapted from Home Experiments at scifun.org. The link is: http://scifun.chem.wisc.edu/HOMEEXPTS/candy.htm

Purpose:

Many different kinds of snack foods and candy are brightly colored, to appeal to our eyes as well as our taste buds. In order to make the candy colored, the manufacturers are allowed to add natural food dyes as well as FDA-certified artificial colors. There are 7 currently approved artificial colors: they are:

- FD&C Blue #1
- FD&C Blue #2
- FD&C Red #3
- FD&C Red #40
- FD&C Green #3
- FD&C Yellow #5
- FD&C Yellow #6

In this experiment we will be using a technique called **chromatography** to look at the colors of M&M's and Skittles and determine if they arise from one or more colored additive. (The word "chromatography" is derived from two Greek words: "chroma" meaning color and "graphein" to write.)

Materials:

- M&M or Skittles candies (1 of each color)
- Rectangle of filter paper
- 600-mL beaker
- water
- table salt
- a pencil (a pen or marker is not good for this experiment)
- a ruler
- 6 toothpicks
- a shallow-well plate.

Procedure:

1. On the filter paper, draw a line with the pencil about ½ inch (1 cm) from one edge of the paper. Make six dots with the pencil equally spaced along the line, leaving at least ¼ inch (0.5 cm) between the first and last dots and the edge of the paper (see diagram below). Below the line, use the pencil to label each dot for the different colors of candy that you have (Y for yellow, G for green, BL for blue, BR for brown, etc.)

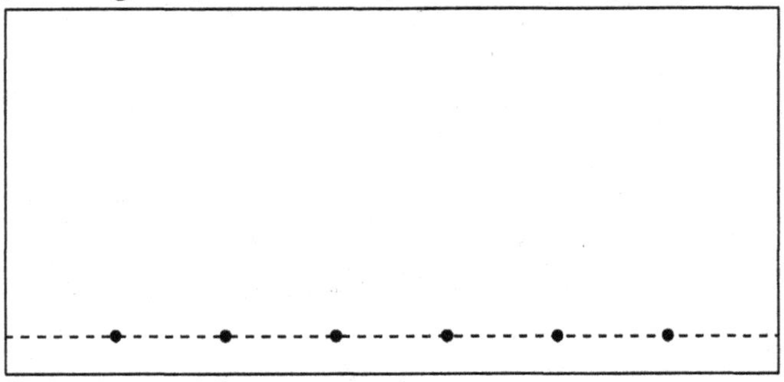

1.

2. Using the shallow-well plate, place one drop of water into 6 different wells. Place one color of candy on each drop. Wait about a minute for the color to come off the candy and dissolve in the water. Remove and dispose of the candies.
3. Touch the tip of one of the toothpicks in one of the colored solutions and lightly touch it to the corresponding dot on the paper. Try to make sure the dot of color stays small. Do the same for the other 5 colors, using a different toothpick each time.
4. After all the spots have dried, repeat step 3 get more color on each spot. Do this a total of three times, waiting for the spots to dry each time.
5. When the paper is dry, roll it so that it forms a loose cylinder which can stand up on its own.
6. In a 400 mL beaker, mix 3.0 grams of salt and 300 mL of water. Stir the contents until all of the salt is dissolved in the water.
7. Now pour some of the solution you made in step 6 into the large 600-mL beaker to a depth of about ¼ inch (0.5 cm). The level of the solution must be low enough so that when you put the filter paper in, the dots will initially be **above** the water level. Set the filter paper into the beaker with the salt solution, and put your watch glass over the top.

Record your observations:

8. When the salt solution has risen to about ½ inch (1 cm) from the top edge of the paper, remove the paper from the solution and mark the paper with a pencil where the solution stopped. Lay the paper on a clean, flat surface to dry.
9. Repeat Steps 1-8 using Skittles instead of M&M's. (and new toothpicks)

* The salt solution is called the mobile phase, and the paper is the stationary phase. We use the word "affinity" to refer to the tendency of the dyes to prefer one phase over the other. The dyes that travel the furthest have more affinity for the salt solution (the mobile phase); the dyes that travel the least have more affinity for the paper (the stationary phase).

Draw (or paste) a representation of your paper chromatography results below:

Calculate R_f values for each of your spots. To calculate the R_f value measure the total distance your substance traveled, from its starting position. Then measure the total distance the solvent traveled, from the starting position of the sample to the solvent front (the furthest point that the solvent reached). Divide the distance your substance traveled by the distance your solvent traveled.

$$R_f = \frac{\text{Distance spot travelled}}{\text{Distance solvent travelled}}$$

Write the R_f value for **each** spot next to it on your diagrams on the previous page. (round off to two decimal places)

Critical Thinking Questions:

1. Compare the spots from the different candies. What are the similarities and differences?

2. Which candies contained mixtures of dyes? Which ones seem to have just one dye?

3. Match some of the colors on the paper with the names of the dyes on the label.

4. Were any of the specific dyes present in more than one color candy? How can you tell? (Hint: look at Rf values).

5. Which color(s) had the most affinity for the salt solution? How do you know?

6. Which color(s) had the least affinity for the salt solution? How do you know?

CHEMISTRY 100: LabActivity 15:
The Sweetness of Aspartame**

GROUP NAMES ________________________ RECORDER ____________

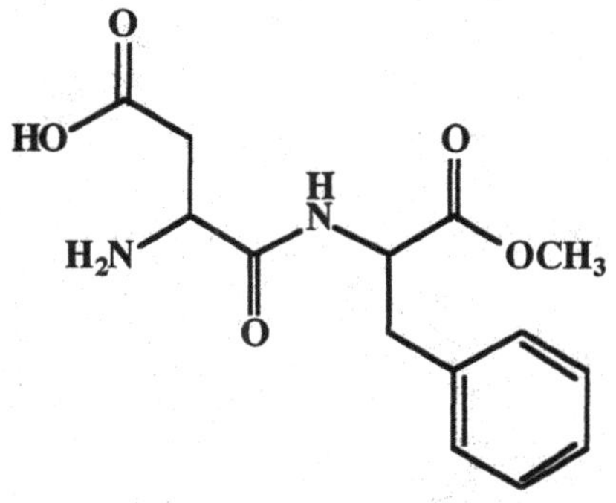

Aspartame (Asp-Phe-OMe)

Background

Aspartame is a non-caloric sweetener used in a broad spectrum of processed food products. It is the methyl ester of the dipeptide aspartyl-phenylalanine. As with other artificial sweeteners, like saccharin and cyclamates, the discovery of its sweetness was serendipitous (by accident) (*1, 2*).

Natural-food advocacy groups (see www.holisticmed.com/aspartame) cite examples of the alleged toxicity of aspartame. There have been instances where consumers linked aspartame consumption with headaches, mood swings, seizures, and other symptoms. Few of these claims are supported by experimental studies, and several pertinent studies are listed here.

- The FDA has established 50 mg/kg/day as the acceptable daily intake (ADI) level for aspartame ingestion (40 mg/kg/day by the World Health Organization), while surveys indicate that the 90th percentile of consumption is between 2 and 10 mg/kg/day (*3*). A single can of diet pop contains 184 mg of aspartame.
- Results of a study involving 120 women taking a single daily dose of approximately 230 mg aspartame indicate no connection between mood swings and aspartame use (*4*). However, this may not be the case in individuals prone to depression, who should avoid excessive aspartame consumption (*5*).
- Between 1986 and 1993, there were 265 reports to the FDA in which aspartame use was believed to be linked to seizures (*6, 7*). Although data are ambiguous at doses exceeding 1,000 mg/kg, studies on individuals with a history of seizures show no evidence for such a connection at lower doses (*8, 9*).

** From Paul J. Stein, in the *Journal of Chemical Education*, **1997**, *74*, 1112

- Diet soda cans contain a warning for people with phenylketonuria. There is some concern that individuals who are heterozygous for phenylketonuria and unaware of this condition might be susceptible to some effects. A review of clinical studies of blood levels of phenylalanine, aspartate, and methanol following aspartame ingestion by phenylketonuric heterozygotes indicates no cause for safety concern by this population (*10*). A study involving five individuals homozygous for phenylketonuria (PKU), ranging in ages from 11 to 23, shows that a single can of diet pop produces only slight increases in plasma phenylalanine levels. These increases are not clinically significant (*11*).

Literature Cited in Background

1. Ellis, J. W. *J. Chem. Educ.* **1995,** *72*, 671–675.
2. Roberts, R. M. *Serendipity: Accidental Discoveries in Science*; Wiley: New York, 1989.
3. Butchko, H. H.; Kotsonis, F. N. *J. Am. Coll. Nutr.* **1991,** *10*, 258–266.
4. Pivonka, E. E. A.; Grunewald, K. K. *J. Am. Diet. Assoc.* **1990,** *90*, 250–254.
5. Walton, R. G.; Hudak, R.; Green-Waite, R. J. *Biol. Psychiatry* **1993,** *34*, 13–17.
6. Maher, T. J.; Wurtman, R. J. *Environ. Health Perspect.* **1987,** *75*, 53–57.
7. Tollefson, L. *Regul. Toxicol. Pharmacol.* **1993,** *18*, 32–43.
8. Tollefson, L.; Barnard, R. J. *J. Am. Diet. Assoc.* **1992,** *92*, 598–601.
9. Sze, P. Y. *Neurochem. Res.* **1989,** *14*, 103–111.
10. Filer, L. J.; Stegink, L. D. *Diabetes Care* **1989,** *12*, 67–74.
11. Mackey, S. A.; Berlin, C. M., Jr. *Clin. Pediatr.* **1992,** *31*, 394–399.

Today we will be determining a dose-response curve to aspartame solutions. Everyone will taste various solutions of aspartame (Equal) in water and rate the sweetness of the sample on a scale from 0-5, comparing the samples to standard solutions.

Procedure

This lab involves tasting of samples. **No glassware or other equipment from your lockers should be used in this experiment**. All of samples will be made in paper or Styrofoam cups. Samples will be tasted using disposable plastic spoons and brand new droppers, and tap water will be provided in jugs. The volume of water will be measured out using a paper cup with a mark on it that has been pre-calibrated to 50 mL.

Today our groups will be combined together to form larger groups with no more than SEVEN students. Each group will be designated by a particular color: Red, Green, or Blue. It will be sufficient to turn in one report for each combined group – write the names of both groups on the first page please.

Sample Preparation

Your group as a whole will be preparing SEVEN solutions of Equal. Each one will contain a different amount of Equal and 50 mL of water. Don't use spatulas from your lab drawer to get the Equal out of the packets: just shake the packets into the paper cups.

Weigh out the following amounts of Equal into seven paper cups:

- 1.000 g
- 0.500 g
- 0.250 g
- 0.125 g
- 0.062 g
- 0.031 g
- 0 g (none)

Add 50 mL of tap water (from the "calibrated" paper cup) to each sample and label them with letters A through G, using a pen with your group's color. DO NOT label them in sequential order: mix them up (maybe the one without Equal is "C", the one with 0.062 g is "F", etc.). A blind taste test won't work if the other group knows which of your samples is which! Decide as a group which one you will label with which letter: make sure you have that information recorded correctly.

Stir your samples with clean plastic spoons to fully dissolve the Equal.

Taste Testing

Your group will obtain seven equal solutions prepared by a different group, and 2 samples labeled "Water" and "1.0 g/50 mL Control". Each group member must taste each sample.

The sweetness of each sample will be rated on a scale from 0 to 5. A 5 means it tastes just as sweet as the 1.0 g/50 mL Control sample, and a 0 means it tastes just as sweet as plain water.

Using a dropper, pipet exactly 5 drops of the 1.0 g/50 mL Control sample onto your plastic spoon and taste it. Mentally make a note of what a sweetness rating of 5 tastes like. Do the same thing with the pure water, and make a mental note of what a sweetness rating of 0 tastes like.

Using your dropper and plastic spoon, taste 5 drops of one of the 7 samples from another group. Rate it on the 0-to-5 sweetness scale and record the value. Have a drink of pure water to cleanse your palate, and taste the next sample. Continue like this until you have tasted every sample from the other group and recorded your sweetness rating.

Once every member of your group has tasted all of the samples, get together and calculate the average of your ratings (to one decimal place). Once you have the average calculated, obtain the masses of Equal in each of the samples you have been tasting from a member of the group which prepared them.

REPORT

Sample Preparation

MY group's color is ____________

Mass of Equal (g) added to 50 mL of water in each solution:

A	B	C	D	E	F	G

Remember: don't reveal this information to the group that is tasting your samples until they have finished computing their averages!

Taste Testing

Group color of samples taste-tested is ____________

Individual sweetness rating of taste testing samples (0 = water, 5 = 1.0 g/50 mL control)

Name	A	B	C	D	E	F	G

Average sweetness rating of taste testing for your group

	A	B	C	D	E	F	G
Average (0.0 – 5.0)							

Mass of Equal (g) used in each solution (obtain this information from the group that prepared the samples that your group tasted):

A	B	C	D	E	F	G

Data Analysis

Draw a graph of the mass of Equal (in each solution you tasted) vs. the group's average sweetness rating.

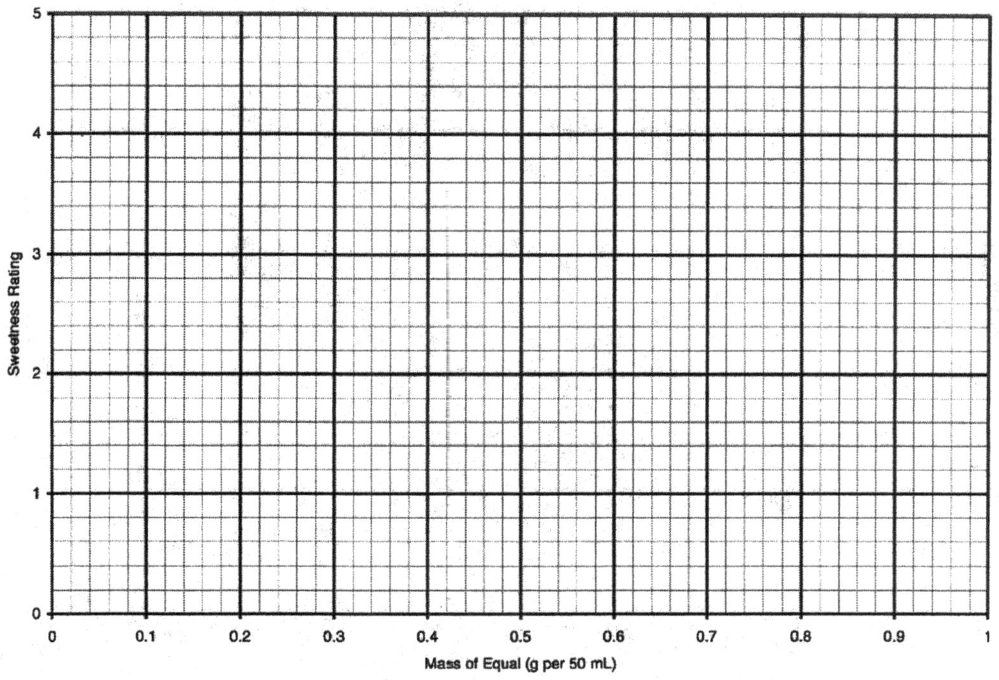

Interpretation

The author of this experiment asserts that, "despite the subjectivity of the taste measurement, the results unequivocally illustrate the hyperbolic nature of the taste response".

1. Is your graph of mass of Equal vs. sweetness linear (increasing evenly) or hyperbolic (leveling off as it goes up)?

2. Based on your results, is the author's statement valid?

3. Can you offer an idea of why a sweetness-taste response should be hyperbolic rather than linear?

For these questions, show your calculations:

4. Dr. Nichols does like his decaffeinated diet sodas. Assuming his mass is 73 kg, what is his maximum acceptable daily intake of aspartame (using the FDA's guidelines of 50 mg/kg/day)?

5. Since each can of soda has about 184 mg of aspartame, what is the maximum number of cans of decaffeinated diet soda that Dr. Nichols can drink before he exceeds the acceptable daily intake? Based on these numbers, should he be concerned about consuming too much aspartame?

6. A one-gram packet of Equal actually contains only 38 mg of aspartame (the rest is filler: glucose and maltodextrose). How many packets of Equal would Dr. Nichols have to consume to exceed his maximum acceptable daily intake?

CHEMISTRY DATA SHEET

PERIODIC TABLE OF THE ELEMENTS

1	2	3	4	5	6	7	8	9	10	11	12	13	14	15	16	17	18
1 H 1.0079																	2 He 4.0026
3 Li 6.941	4 Be 9.0122											5 B 10.81	6 C 12.011	7 N 14.007	8 O 15.999	9 F 18.998	10 Ne 20.179
11 Na 22.99	12 Mg 24.305											13 Al 26.982	14 Si 28.086	15 P 30.977	16 S 32.06	17 Cl 35.453	18 Ar 39.948
19 K 39.098	20 Ca 40.08	21 Sc 44.956	22 Ti 47.88	23 V 50.942	24 Cr 51.996	25 Mn 54.938	26 Fe 55.847	27 Co 58.933	28 Ni 58.69	29 Cu 63.546	30 Zn 65.38	31 Ga 69.72	32 Ge 72.59	33 As 74.922	34 Se 78.96	35 Br 79.904	36 Kr 83.80
37 Rb 85.468	38 Sr 87.62	39 Y 88.906	40 Zr 91.22	41 Nb 92.906	42 Mo 95.94	43 Tc (98)	44 Ru 101.07	45 Rh 102.91	46 Pd 106.42	47 Ag 107.87	48 Cd 112.41	49 In 114.82	50 Sn 118.69	51 Sb 121.75	52 Te 127.60	53 I 126.9	54 Xe 131.29
55 Cs 132.91	56 Ba 137.33	57 La 138.91	72 Hf 178.49	73 Ta 180.95	74 W 183.85	75 Re 186.21	76 Os 190.2	77 Ir 192.22	78 Pt 195.08	79 Au 196.97	80 Hg 200.59	81 Tl 204.38	82 Pb 207.2	83 Bi 208.98	84 Po (209)	85 At (210)	86 Rn (222)
87 Fr (223)	88 Ra 226.03	89 Ac 227.03	104 Rf (261)	105 Db (262)	106 Sg (263)	107 Bh (262)	108 Hs (265)	109 Mt (266)	110 Uun (269)								

58 Ce 140.12	59 Pr 140.91	60 Nd 144.24	61 Pm (145)	62 Sm 150.36	63 Eu 151.96	64 Gd 157.25	65 Tb 158.93	66 Dy 162.50	67 Ho 164.93	68 Er 167.26	69 Tm 168.93	70 Yb 173.04	71 Lu 174.97
90 Th 232.04	91 Pa 231.04	92 U 238.03	93 Np 237.05	94 Pu (244)	95 Am (243)	96 Cm (247)	97 Bk (247)	98 Cf (251)	99 Es (254)	100 Fm (257)	101 Md (258)	102 No (259)	103 Lr (262)

FORMULAS

$E = h\nu$ $\quad c = \lambda\nu$

$\frac{1}{\lambda} = R\left(\frac{1}{m^2} - \frac{1}{n^2}\right)$

$w = -P\Delta V$

$\Delta G = \Delta H - T\Delta S$

$q = mc\Delta T$

$\bar{v} = \sqrt{\frac{3RT}{M}}$

$PV = nRT$

$\left(P + \frac{an^2}{V^2}\right)(V - nb) = nRT$

$\ln P_v = \left(\frac{-\Delta H_{vap}}{R}\right)\frac{1}{T} + C$

$\Delta T_b = K_b m$ $\quad \Delta T_f = K_f m$

$p_i = X_i P_{total}$

$\Pi = MRT$

$\ln [A] = -kt + \ln [A]_0$

$\frac{1}{[A]} = kt + \frac{1}{[A]_0}$

$t_{\frac{1}{2}} = \frac{0.693}{k}$

$k = Ae^{-E_a/RT}$

$\ln\left(\frac{k_2}{k_1}\right) = \frac{E_a}{R}\left(\frac{1}{T_1} - \frac{1}{T_2}\right)$

$x = \frac{-b \pm \sqrt{b^2 - 4ac}}{2a}$

CONSTANTS

$N_A = 6.022 \times 10^{23}\ mol^{-1}$

R (Rydberg)$=1.097 \times 10^{-2}\ nm^{-1}$

$c = 2.998 \times 10^8\ m/s$

$h = 6.626 \times 10^{-34}\ J \cdot s$

$m_e = 9.11 \times 10^{-31}\ kg$

$e = 1.602 \times 10^{-19}\ C$

$R = 0.08206\ \frac{L \cdot atm}{mol \cdot K}$

$R = 8.314\ \frac{J}{mol \cdot K}$

$R = 62.37\ \frac{L \cdot mmHg}{mol \cdot K}$

CONVERSION FACTORS

1 cal = 4.184 J

1 atm = 101.325 kPa = 760 mmHg

0 °C = 273.15 K

1 L·atm = 101.325 J

1 inch = 2.54 cm

1 mile = 1.609344 km

1 lb = 453.6 g

1 Å = 10^{-10} m

1 mL = 1 cm^3

STP = 1 atm, 0 °C.

www.ingramcontent.com/pod-product-compliance
Ingram Content Group UK Ltd.
Pitfield, Milton Keynes, MK11 3LW, UK
UKHW050722170926
15229UKWH00010B/45

9 781524 926847